T0262760

URBAN ENVIRONMENTAL PLANNING

Urban Environmental Planning

Policies, Instruments and Methods in
an International Perspective

Edited by
DONALD MILLER
GERT DE ROO

Routledge
Taylor & Francis Group

LONDON AND NEW YORK

First published 2005 by Ashgate Publishing

Published 2016 by Routledge
2 Park Square, Milton Park, Abingdon, Oxon OX14 4RN
711 Third Avenue, New York, NY 10017, USA

Routledge is an imprint of the Taylor & Francis Group, an informa business

British Library Cataloguing in Publication Data
Urban environmental planning : policies, instruments and
    methods in an international perspective. - 2nd ed. - (Urban
    planning and environment)
    1. City planning - Environmental aspects 2. Environmental
    policy
    I. Miller, Donald II. Roo, Gert de
    307.1'216

Library of Congress Cataloging-in-Publication Data
Urban environmental planning : policies, instruments and methods in an international
perspective / [edited] by Donald Miller and Gert de Roo.
        p. cm. -- (Urban planning and environment)
    Originally published: Aldershot, Hants, England ; Brookfield, Vt. : Avebury, c1997.
"Papers presented at an international symposium on urban planning and environment,
held in Seattle, USA, in March 1994"--Introd.
    Includes bibliographical references and index.
    ISBN 0-7546-4392-1
        1. City planning--Environmental aspects. 2. Land use, Urban--Environmental aspects.
    3. Urban ecology. 4. Environmental protection. I. Miller, Donald. II. Roo, Gert de.
    III. Series.

    HT166.U727 2004
    307.1'216--dc22

                                                                                    2004015405

ISBN 9780754643920 (hbk)

Contents

List of Figures

List of Tables

Annexes

List of Contributors

N. Anderson, Ph.D., New York City Department of Environmental Protection, Community Environmental Development Group.

D. Arbouw, Senior Advisor Environmental Affairs, Province Noord-Holland, Haarlem, The Netherlands.

N. Border, Faculty of the Built Environment, University of the West of England, Bristol, England.

H. Borst, Directorate General for Spatial Planning, Dutch Ministry of Housing, Spatial Planning and Environment.

G.M. Bush, Planning and Real Estate Division, Metro, Seattle.

P. Chung Chan, Government Town Planning, Planning Department, Hong Kong.

J. Denkers, Washington Department of Ecology, Nuclear Waste Program.

J. Farinha, Assistant Professor, Department of Sciences and Environmental Engineering, New University of Lisbon, Portugal.

C. Garrett, General Director of the Ambiente of Lisbon, Portugal.

B. Goodall, Professor and Head of the Department of Geography, University of Reading, England.

E. Hanhardt, New York City Department of Environmental Protection, Community Environmental Development Group.

C. Marks, Seattle Planning Department, USA.

D. McNamara, King County Environmental Division, Bellevue, Washington, USA.

E. Meijburg, Amsterdam Department of Environmental Affairs, Amsterdam, The Netherlands.

D. Miller, Professor, Department of Urban Design and Planning, University of Washington, Seattle, USA.

E. Munday, Noise Remedy Manager, Port of Seattle, Sea-Tac, Washington, USA.

S. Nicholas, Senior Environmental Planner and co-ordinator of the City's Environmental Priorities Project, Seattle Office of Management and Planning.

I. Pasher, Community District #1, Brooklyn, New York.

J. Pearce, Freelance Advisor and Researcher, The Dutch Ministry of Housing, Spatial Planning and Environment.

A. Perestrelo, Lecturer, Department of Sciences and Environmental Engineering, New University of Lisbon, Portugal.

M. Power, Washington Department of Ecology, Nuclear Waste Program.

K.O. Richter, King County Environmental Division, Bellevue, Washington, USA.

G. de Roo, Professor, Faculty of Spatial Sciences, University of Groningen, The Netherlands.

K.J. Stenberg, King County Environmental Division, Bellevue, Washington, USA.

N. Streefkerk, Department of Environmental Affairs, Municipality of Zwolle, The Netherlands.

E. Timár, Environmental Department of the City of Amsterdam, The Netherlands.

L. Vasconcelos, Lecturer, Department of Sciences and Environmental Engineering, New University of Lisbon, Portugal.

L. Vicknair, King County Environmental Division, Bellevue, Washington, USA.

F. Westerlund, Associate Professor, Department of Urban Design and Planning, University of Washington, Seattle.

M.A. Wolf, Professor of Law and History, American University and University of Richmond, USA.

Preface

Integrated environmental planning, smart growth planning, growth management planning: by whatever name, contemporary city planning in societies around the world is adopting sustainable urban development as its over-arching goal. Sustainability seeks to balance social concerns with economic and environmental concerns. It is now a challenge to planning programs to discover and apply creditable strategies for making this goal operational.

The major purpose of this book is to share what has been learned from a number of initiatives to accomplish this. These initiatives have been undertaken by local and regional governments throughout the world, with emphasis on The Netherlands and the US.

These cases are reported and assessed by people who were directly involved and well informed about them. For this reason, these cases provide ideas and experiences that have direct value to public officials seeking to deal with urban environmental issues through new forms of city planning. A major reason for publishing this second edition is to make it easier to use in university courses addressing urban environmental planning, since the cases presented in it serve as useful demonstrations of new planning approaches and methodologies.

This book is one in a series published by Ashgate, that supports the mission of the International Urban Planning and Environment Association: to foster exchanges of information concerning new and promising practices in planning that seek to improve environmental quality and sustainable development. The editors greatly appreciate the support of the staff at Ashgate Publishers in making this series possible and a success. We also owe special thanks to Floris Bruil for reformatting the text for this edition, and to Tamara Kaspers and Johan Zwart for redesigning the figures used in the text.

Donald Miller and Gert de Roo
Seattle, USA and Groningen, The Netherlands

International Urban Planning and Environment Association

Chapter 1

Introduction

D. Miller[1] and G. de Roo[2]

1.1 Improving Environmental Quality in Cities

Making cities environmentally healthful places in which to live, work and play is a growing concern throughout the world. The focus of much of this concern is on ways to measure and manage the environmental spillovers of one urban function on other urban activities. Some of the most dangerous and annoying spillovers result from manufacturing activities and transport. These impacts are most acute for residential areas and other environmentally sensitive urban functions.

In the past, city planning has sought to reduce or eliminate these impacts by physically separating environmentally intrusive and environmentally sensitive land uses. Elected officials and planners have come to realise that this strategy is no longer effective as cities grow in size, and as local and national policies call for increasing the density and mixture of uses in urban areas.

This book consists of a collection of papers reporting on major programs in a number of countries to address these issues. These contributions are selected from more than sixty papers presented at an international symposium on Urban Planning and Environment, held in Seattle, USA, in March 1994. This symposium was sponsored by the Ministry of Housing, Spatial Planning and Environment of The Netherlands. It was organised by the Faculty of Spatial Sciences at the University of Groningen, The Netherlands, and by the Department of Urban Design and Planning at the University of Washington in Seattle. The four days of prepared presentations and deliberations provided an exceptional opportunity for governmental officials, researchers in the field, and citizen representatives from thirteen countries to explore mutual problems and to consider the possible applicability of programs from other cities and societies as solutions to environmental quality problems that they are dealing with in their own contexts.

A major purpose of this symposium was to search for solutions to the problems of negative environmental spillovers from urban activities on other urban land uses, including the effects of noise, odour, vibration, risks from explosion and fire, and toxic and carcinogenic substances in air pollution. Soil and water pollution are also addressed by some of these papers. An additional purpose was to explore means of fostering positive environmental spillovers, produced by land uses such as wetlands and parks. This is arguably the most critical important set of topics in planning and developing sustainable cities (Breheny 1992).

The initiative for holding this international symposium came from the Dutch government, which wanted to share ideas and experiences from their development of Integrated Environmental Zoning and to get critical responses to this program. Additionally, Dutch officials wished to learn about counterpart programs underway in other countries, and about experiences with these.

Several of the papers selected for this volume describe and critically assess experiments with applying this innovative Dutch program, and variants of this approach being applied in Amsterdam, as discussed by Meyburg and by Arbouw. Most of the papers similarly present cases of public sector programs to measure and manage environmental spillovers within urban areas of other countries. Finally, several of the papers report on research which is supporting these efforts. Together, these papers represent a rich resource of ideas concerning methodologies and approaches for addressing these issues, and additionally discuss the political, institutional and economic aspects of implementing these programs. Lessons from the concrete cases which are presented have practical value in guiding the design of programs to improve urban environmental quality in other contexts.

1.2   Environmental Conflicts within Urban Areas

Environmental protection legislation adopted in a number of countries during the 1970s increased attention to reducing the negative effects of human activities. This legislation also resulted in increased awareness of the quality of our surroundings in cities, and brought into focus differences in the approaches of physical and environmental planning.

Physical or spatial planning has sought answers to questions such as 'Where can we locate this activity?' and 'What can be located here?' Environmental planning takes an opposite perspective, asking 'Where can this activity not be located?' and 'What can not be located here?' The differences in these perspectives reveal a conflict in principles concerning the use of space in cities.

Spatial planning seeks to locate urban activities in a way which most benefits society. Lately this has involved seeking to integrate a variety of functions within the same area in order to reduce trips to work and other activities and to reduce pressures for development of remaining open-use lands especially at the edges of urban areas. However, locating activities with significant negative environmental impacts too near to environmentally sensitive activities may reduce the environmental quality enjoyed by residents.

Environmental planning, on the other hand, seeks to improve or protect environmental quality for residents, both through controlling the generation of pollution and through segregating activities that are environmentally incompatible. This physical separation can work against spatial plans aimed at reducing trip making through mixing land uses and increasing densities: a major proposal of the European Commission Green Paper on the Urban Environment (CEC 1990) and of the 'compact city' policy which is increasingly popular throughout the world (Owens 1992; Spiller 1993).

1.3   Integrated Environmental Zoning

The Dutch initiative to assess and improve environmental quality in cities is one of the few efforts to integrate these concerns and to resolve this conflict. As several of the following contributions point up, it is one of the still fewer innovative approaches which is being employed in a number of pilot projects to provide the basis for assessing how well it works and for identifying needed improvements. Some of these pilot projects have shown that the threat to health and well being of existing pollution levels are much greater than had previously been supposed, and that the public and private sector costs of correcting these situations are greater than currently available resources will support. An overview of this program serves to introduce papers which provide more detailed description and assessment in following sections of this book.

The Provisional System for Integrated Environmental Zoning (IEZ) was initiated in 1990 and is based on assessing several kinds of spillovers from industry or traffic on areas occupied by sensitive uses (VROM 1990a). Standards for nuisances and threats to safety and health, such as 50 dB(A) for noise, are used to designate what is acceptable exposure in residential areas and what is not. These standards are used to map zones which indicate which parts of an urban area are heavily impacted by these negative spillovers. Environmental quality problems are identified when portions of a city mapped as heavily impacted are occupied by sensitive activities. In these cases, efforts are made to reduce pollution at the source and, if for whatever reasons this is not possible, residents are moved from areas of intense pollution and the housing is removed (VROM 1990b). When sensitive uses are not located in these polluted areas, the local land use plan can restrict future residential development within these areas.

Reports on programs in other parts of the world which are included in this book demonstrate that spatial assessment of environmental impacts is not new. However the Dutch IEZ program is innovative in assessing the additive effects of noise, air pollution and industrial hazards. Thus, in residential areas impacted both by odour and noise will receive a lower environmental quality rating than if these effects were measured individually.

Arnhem, a city of about two hundred thousand inhabitants, is the location of one of the eleven pilot projects and serves as an illustration of results from applying the provisional system of IEZ. Manufacturing firms in the industrial area Arnhem Noord are polluting a portion of that city about eight kilometres square so intensely that it has been designated a 'black area': a category with combined pollution levels unacceptable for residential use. The spillovers are H_2S which produces an unpleasant odour, and CS_2 which is toxic. Restricting unacceptable levels of pollution to the industrial site in Arnhem is virtually impossible within a few years' time. In this situation, the IEZ program calls for shutting down factories employing thousands of people, or vacating major residential areas of Arnhem (Boli 1993). Several of the other pilot projects have posed similar dilemmas (de Roo 1993).

While Integrated Environmental Zoning is only one method for assessing the incidence and effects of pollution in cities (Cappon 1990; Lee and Walsh 1992), it is unfortunately one of the few comprehensive methods which is currently being tested in practice (Ries and Roseland 1991). The more common procedure is to collect and analyse information concerning pollution effects in a sectoral manner – one type of impact at a time – which fails to account for the environmental problems posed by several of these spillovers in combination (OECD 1990). As several of the following papers point out, integration of spatial planning and environmental policy is a promising approach to relating the locational and spatial demands of different urban functions to environmental quality concerns.

1.4   Major Issues in Designing Programs to Improve Environmental Quality in Cities

The recent emergence of urban environmental planning, and the discussions which took place at the 1994 symposium in Seattle, are not limited to making a choice between a sectoral or an integrated approach to dealing with various forms of pollution and their impacts. Presentations at the Seattle symposium reported on efforts that many countries are employing to resolve environmental conflicts in urban areas, each in their own way and influenced by the planning history of each country. However these presentations and ensuing discussions raised a number of themes and kinds of issues which cut across national differences and are seen as common to efforts to improve urban environmental quality. We have chosen to address these issues in terms of distinctions, some of which are original and others are new applications of earlier concepts. Each of these distinctions provide a basis for evaluating existing programs and to assess new proposals. Additionally, these distinctions provide a framework for viewing the contributions included in this volume. The following are several of these distinctions, a brief definition of each of them, and a comment on their significance.

Negative versus Positive Spillovers

Negative spillovers (or neighbourhood effects, or externalities) are undesirable impacts generated by the behaviour of one party on other parties, for which the impacted parties are not compensated, as in the case of air or water pollution. This concept is widely used by environmental quality programs and, being a non-market effect, is the basis of public sector regulation. Positive spillovers, on the other hand, refer to desirable effects generated by one party and enjoyed by other parties, such as the amenity provided by parks and enjoyed by surrounding residents. Positive externalities have not been widely recognised by environmental quality programs, yet they importantly contribute to quality of life, and can in some cases be designed as an element of a mitigation program to offset some negative effects, as for example discussed by Meyburg in a following section of this book. Since they are non-market goods, positive neighbourhood effects will seldom be provided by the private sector unless they are required by the public sector as part of a comprehensive mitigation program or are publically subsidised, as illustrated by the papers by Marks, by Bush, and by Stenberg and others.

Top-Down versus Bottom-Up Organisational Responsibility

This distinction focuses on whether subordinate or superordinate levels of government have the major responsibility for designing environmental quality improvement policies and programs, to set standards, and to control implementation. In most cases presented at the Seattle symposium, national governments directed local governments to carry out programs and to employ methods set at the higher level to assure that all members of society enjoy a healthy and desirable environment. However, a number of the papers that follow identify difficulties which local governments have in taking locally appropriate action because of the lack of flexibility allowed by the central government. This has resulted in reducing creativity at the local level, inefficient use of resources including overly costly mitigation programs, and threats to the competitiveness of local firms with consequences for economic development. An important area of inquiry is to seek an effective balance between the authority of the local and central levels of government.

Source versus Effect Oriented Strategies

Source oriented strategies seek to reduce pollution and other hazards by regulating the performance of those activities, such as manufacturing, which are generating environmental impacts. This is the conventional approach used in environmental management programs, and is the simplest means to place responsibility on the polluter, that is to cause firms to internalise their externalities. Effect oriented strategies seek either to shield environmentally sensitive activities from impacts, such as insulating housing from highway or airport noise, or to spatially separate sensitive activities from sources of pollution. Land use planning in the past has primarily focused on effect oriented strategies, often using distance to mitigate externalities which can not be limited to the site at which they are generated. Some of the more innovative recent programs seek to combine both of these strategies in an appropriate manner, thus merging the contributions of environmental management and land use planning.

Sophistication versus Practical Applicability

This is not a new distinction, but an important consideration and sometimes poses a dilemma in designing an effective strategy for improving environmental quality in urban areas. While it is important that both the analysis and program design be well informed, that is that it be both reliable and valid, evidence from some of the contributions in this book point out that methodological sophistication sometimes results in very expensive data collection and analysis, and in analytical frameworks that only experts understand and appreciate, making these schemes difficult to apply. Effective environmental improvement programs require a balance between information richness and simplicity, and we currently know little about what this balance should be.

Market Based versus Regulatory Approaches

These two categories of methods for implementing programs to improve environmental quality are widely known. However, as Goodall's paper points out, regulation is the most commonly employed method. Market based methods seek economic incentives for desired behaviour, such as reducing pollution emissions. These inducements are an effort to 'pull' parties into compliance, and may include a range of tactics from public recognition and improved good will among consumers and others, to tax reduction and other financial incentives, in a manner similar to the use of bonus zoning. In contrast, regulations stipulate required performance, usually enforced through fines or other punitive action. This enforcement involves 'pushing' the parties to comply with, for example, emission standards. Discussion at the symposium noted the greater psychological acceptance of inducements than of regulations, but also noted the difficulty in developing effective market based approaches. One example of this are experiments with issuing credits to plants that reduce pollution which, when pollution abatement exceeds required emission reductions, may be traded by that firm to another firm to use in partially meeting its emission reduction target, at less cost than purchasing the technology to do so. The discussion recognised that inducements and regulations are best seen as complements to each other, and that use of emission reduction credits is an example of this. Research to identify and test additional market based approaches is needed and their application could have significant value not only in encouraging intrusive activities to improve their performance, but in developing wider political acceptance of programs to improve environmental quality in cities.

Technical versus Perceptual Methods

Technical methods rely on scientific knowledge, and include measuring the levels of environmental impact on a neighbourhood as well as translating these impacts into their effects on the health and well being of the residents. Perceptual methods involve how people regard these impacts and how much they are distressed by them. While the former implies objective reasoning, the latter recognises the social and psychological importance of what is often regarded as subjective valuation. This distinction is addressed by Anderson and others in the case of a program in Brooklyn, and by Nicholas concerning risk-based analysis in Seattle.

Perceptual concerns are especially important in instances in which impacts are a source of annoyance and lowered quality of life, while technical methods may be most appropriate where impacts are health threatening and may be inconspicuous to the casual observer. While more work needs to be done on this issue, impacts that cause annoyance may be negotiable in environmental programs, while regulation of health or life threatening effects should be non-negotiable.

Strategic versus Operational Approaches

A widely recognised problem in planning both in the public and private sectors is how to link abstract goals and policy with practical measures to implement these ends.

Strategic planning (Steiner 1979; Bryson and Roering 1988; So 1986; Gordon 1993) is a recent development which employs a process for designing goals and mission statements, and integrating this activity with tactical planning which includes operational management activities. The value of applying this integrative approach for dealing with 'big picture' issues and related specific actions and programs to improve and protect urban environmental quality appears to be promising, although it has not yet been tested by research or widely by practice. Some contributions, including those by Nicholas and by Arbouw, discuss beginning efforts to employ approaches combining strategic planning and operational management, and suggest the value of additional experimentation with this model.

Participation versus Prescription

Agenda 21 (Quarrie 1992), the Green Paper on the Urban Environment (CEC 1990), and most other recent international declarations on environmental issues recognise the need and value of participation in defining problems and finding solutions to them. Participation is facilitated by decentralising decision making to the local level to the extent possible, and includes all impacted parties: all residents, managers of intrusive activities, politicians, and those responsible for managing the implementation of programs. The purposes served by participation include the principle of democracy, that people with immediate experience with the causes and effects of pollution have information that is invaluable to designing an effective program, and that participation not only builds knowledge and competence on the part of those participating but develops a political constituency for the program. These are the reasons that participation and a bottom-up approach are major principles in strategic planning, discussed earlier.

Prescription, on the other hand, is used to convey that standards and procedures are technically determined and specified as required features of programs, as illustrated in the contribution by Borst. Not uncommonly, prescription appears to be less a function of better information based on scientific analysis than a function of conserving and enhancing the power of those doing the prescribing. Sometimes it is also a function of wanting to avoid the complications and time involved if the process is participatory.

While there is similarity between this distinction and our earlier discussion of technical versus perceptual methods, this distinction deals with the planning process and who is involved in it. Discussion at the symposium concluded that both participation and prescription have roles to play, especially if the latter is for valid reasons. These reasons may be valid because of scientific evidence such as that some environmental effects are toxic but are not apparent, and even because locally generated effects impact areas and people outside of the locality, as is the case with acid rain and global warming. Additionally, criteria and procedures prescribed by the central government may be an effective way of forcing local parties to confront difficult political decisions which they may otherwise seek to avoid.

As with several other distinctions, discussions at the symposium concluded that prescription and participation can be useful complements to each other in designing a program to improve urban environmental quality, but that we need to develop improved models and means of participation for use to this end. It is one thing to espouse participation, as many general statements have, and quite another thing to design and implement a program for a specific context that successfully accomplishes it, as for example the contribution by Chan points up. Critical investigation of a range of methods that have been tried, and of new methods, could be a valuable resource for designing a participation program appropriate to a local situation and set of issues.

Integration versus Separation of Urban Activities

Spatial or physical planning has emphasised a strategy of minimising the effects of environmental spillovers from one land use on other land uses by separating these activities within urban areas, thus using distance as a means of reducing these effects. More recently, and especially with increasing interest in the compact city concept, planners have sought ways of spatially integrating various urban functions, or at least bringing them much nearer to each other. The major purpose of this strategy is to reduce the number and length of trips, not only from home to work but for shopping and other activities, and to encourage use of public transit and walking instead of the private automobile (Jones 1993).

A number of the presentations at the symposium in Seattle reported on local efforts to encourage this pattern of development and thus to reduce air and water pollution, as well as to diminish traffic congestion. Questions are now being raised whether this urban form has much effect on reducing automobile trips, and whether other programs are not more cost-effective in accomplishing this objective (Downs 1992; Brotchie 1992; Breheny and Rockwood 1993).

As integration of urban land uses is considered by various localities, urban environmental planning will need to assess the evidence on whether and how much this approach will result in reduced vehicle emissions. It will also need to assess whether and how much environmentally sensitive activities such as housing and recreation will be more greatly impacted by the spillovers from intrusive activities which will now be placed in greater physical proximity to them.

1.5   Conclusions

While several additional distinctions emerged from the discussions and papers at the symposium in Seattle, these nine are among the most important and illustrate some of the contributions made by the papers presented in this volume. Each of these issues will be worth pursuing further because their resolution is important to providing the understanding needed in designing effective programs of analysis and action aimed at improving environmental quality in cities throughout the world. If these issues are not resolved, many initiatives are likely to be wasteful and misdirected.

Residents are demanding greater liveability in their city, and quality of life is seen as a major source of competitive advantage in attracting business investment (Power 1980; Schmenner 1982). Additionally, sustainability and planning for sustainable urban development have recently become major issues receiving world wide attention (Haughton and Hunter 1994; CEC 1992). The emergence of these concerns increases the demand for effective urban environmental planning, a field that is still being defined. As we have suggested, to be effective it will need to include many of the concerns and much of the content of both the spatial planning and environmental management practices of the last few decades.

Papers and discussions at the symposium in Seattle confirmed that most countries are only just beginning to address the problem of assessing and improving urban environmental quality in a comprehensive manner, and fewer are experimenting with possible solutions. The contributions presented in this book provide many useful ideas, not only in the form of analytical methodology but in terms of political requirements and organisational strategies which are important features of an effective program. The case studies ground most of these contributions in practice, providing concrete illustrations of programs that succeed and those that are still seeking success.

Notes

1   Donald Miller is Professor at the Department of Urban Design and Planning, University of Washington, Seattle, USA.

2   Gert de Roo is Professor in Planning at the Faculty of Spatial Sciences, University of Groningen, Groningen, The Netherlands.

References

Boei, P.J. (1993) Integrale Milieuzonering op en rond het Industrieterrein Arnhem-Noord [Integrated Environmental Zoning on and around industrial site Arnhem Noord], in G. de Roo (ed.) *Kwaliteit van Norm en Zone* [Standard and Zone Quality], University of Groningen, Groningen, The Netherlands, pp. 75-82.

Breheny, M.J. (1992) Towards Sustainable Urban Development, in S.R. Bowlby and A.M. Mannion (eds), *Environmental Issues in the 1990s*, Wiley and Sons, Chichester.

Breheny, M.J., and R. Rockwood (1993) Planning the Sustainable City Region, in A. Blowers (ed.) *Planning for a Sustainable Environment*, Earthscan, London.

Brotchie, J. (1992) The Changing Structure of Cities, *Urban Futures*, Vol. 5, pp. 13-23.

Bryson, J.M., and W.D. Roering (1988) Applying Private Sector Strategic Planning in the Public Sector, in J.M. Bryson and R.C. Einsweiler (eds) *Strategic Planning: Threats and opportunities for planners*, Planners Press, Chicago, pp. 15-34.

Cappon, D. (1990) Indicators for a Healthy City, *Environmental Management and Health*, Vol. 1, 1, pp. 9-18.

CEC (Commission of the European Communities) (1990) Green Paper on the Urban Environment, EEC, Brussels.

CEC (Commission of the European Communities) (1992) Towards Sustainability: A European Community programme of policy and action in relation to the environment and sustainable development, EEC, Brussels.

Downs, A. (1992) *Stuck in Traffic*, The Brookings Institution, Washington, DC.

Gordon, G.L. (1993) *Strategic Planning for Local Government*, International City Management Association, Washington, DC.

Haughton, G., and C. Hunter (1994) *Sustainable Cities*, Jessica Kingsley Publishers, London.

Jones, G. (1993) Planning and the Reduction of Transport Emissions, *The Planner*, Vol. 77, 7, pp. 15-20.

Lee, N., and F. Walsh (1992) Strategic Environmental Assessment: An overview, *Project Appraisal*, 7:3, pp. 126-136.

OECD (1990) Environmental Policies for Cities in the 1990s, OECD, Paris.

Owens, S. (1992) *Energy, Environmental Sustainability and Land Use Planning*, in M. Breheny (ed.) *Sustainable Development and Urban Form*, Pion, London.

Power, T.M. (1980) *The Economic Value of the Quality of Life*, Westview Press, Boulder, CO.

Quarrie, J. (ed.) (1992) *Earth Summit 1992: The United Nations conference on environment and development*, Regency Press, London.

Rees, W., and M. Roseland (1991) Sustainable Communities: Planning for the 21st century, *Plan Canada*, Vol. 31, 3, pp. 15-26.

Roo, G. de (1993) Environmental Zoning: The Dutch struggle towards integration, *European Planning Studies*, Vol. 1, 3, pp. 367-377.

Schmenner, R. (1982) *Making Business Location Decisions*, Prentice Hall, Englewood Cliffs, NJ.

So, F.S. (1986) Planning Agency Management, in F.S. So et al. (ed.), *The Practice of Local Government Planning*, Second Edition, International City Management Association, Washington, DC, pp. 401-434.

Spiller, M. (1993) Federal Initiatives for Better Cities, *Urban Futures*, Vol. 3, 1, pp. 15-24.

Steiner, G.A. (1979) *Strategic Planning: What every manager must know*, Free Press, New York, NY.

VROM [Dutch Ministry of Housing, Spatial Planning and Environment] (1990a) Aktieplan Gebiedsgericht Milieubeleid, Tweede Kamer 1990/1991, 21896, No. 1-2, SDU, The Hague.

VROM. [Dutch Ministry of Housing, Spatial Planning and Environment] (1990b) Integrale Milieuzonering deel 14 [Ministerial Manual for a Provisional System of Integral Environmental Zoning], VROM, Leidschendam, The Netherlands.

Part A
A Descriptive Introduction to Environmental and Spatial Conflicts in the Urban Area

Chapter 2

Pollution Control in the United Kingdom

B. Goodall[1]

2.1 Pollution Control and Environmental Policy

Pollution control must be an integral component of environmental policy, particularly where the latter involves commitment to a sustainable development strategy. Environmental pollution control in the United Kingdom (UK) has evolved over the past 150 years, increasing in coverage and sophistication as scientific understanding of pollution impacts became more acute and as society accorded environmental quality a higher priority. For most of those 150 years, the UK approach to pollution control has focused on individual environmental media, i.e. air, water or land, and has been based on a philosophy of 'command and control', or regulation. The emphasis has been on developing policies to protect the environment against pollution from industry and other sources and these longstanding initiatives have been revised regularly in the face of changing circumstances and newly-perceived threats. Likewise the administrative and enforcement procedures have been updated to improve the effectiveness of these pollution control policies. However, responsibility for pollution control was horizontally fragmented and was characterised by a lack of integration with general environmental planning and waste management policies.

A marked change of attitude is discernible from the late 1980s on a number of counts. Firstly is realisation of the limited effectiveness of pollution control legislation functioning independently of planning. Control of pollution needed to be viewed as part of a comprehensive approach to the protection of the environment. Planning has been the principal tool of environmental protection in the UK and, in recent years, increasing awareness of environmental priorities has led planning authorities to take a greater interest in controlling potentially polluting activities. However, planning controls are not viewed as an appropriate means of regulating the detailed characteristics of potentially polluting activities (Department of the Environment 1994). Moreover, planning is largely impotent when it comes to imposing further restrictions on established industry. The planning and pollution control systems operating in the UK may be separate but they should be complementary.

Secondly is the realisation that pollution frequently transcended national boundaries and that increasingly action needed to be undertaken on an international basis to protect the environment against, for example, ozone depletion, marine pollution, and the threat of climate change. The UK is therefore party to various international initiatives and agreements, for example, the *Montreal Protocol – the*

1987 Agreement on Substances which Deplete the Ozone Layer under which the supply of chlorofluorocarbons (CFCs), halons and carbon tetrachloride was originally to be phased out by the year 2000 and of 1,1,1 trichloroethane by 2005 (but with tighter timetables agreed under the Protocol's 1990 and 1992 revisions) (Central Office of Information 1993). In this wider context the UK Government has published a series of White Papers bringing together a range of policies for the protection of the environment. The Environment White Paper, *This Common Inheritance* (UK Government 1990), was the first comprehensive statement by the British Government of its policy on issues affecting the environment and clearly demonstrated awareness of concerns over sustainable development raised by the earlier Brundtland Report, *Our Common Future* (World Commission on Environment and Development 1987). Likewise the UK Government has responded to the four major areas of agreement at the 'Rio' or 'Earth Summit', the 1992 United Nations Conference on Environment and Development:

- the Agenda 21 call for comprehensive programs for achieving sustainable development;
- the Climate Change Convention requiring action to reduce global warming;
- the Biodiversity Convention requiring protection of species and habitats;
- the need for a statement of principles for the management of the world's forests.

These responses are set out in a series of strategy documents (UK Government 1994a; 1994b; 1994c; 1994d). Changes to UK pollution control policy must be viewed in this wider context.

Thirdly is realisation that pollution control legislation had limited effectiveness where restricted to individual, constituent parts of the environment. This change of heart is consistent with that in the wider European Community where the advantages of a holistic approach to pollution control are also recognised. 1990 represents a high-water mark in UK legislative thinking on environmental matters since alongside the Environment White Paper, *This Common Inheritance*, is the *1990 Environmental Protection Act*, which establishes the fundamentals of the new approach referred to as *Integrated Pollution Control* (IPC), which is consistent with the *Polluter Pays Principle* (PPP), and additionally an emphasis on preventive control of waste and emissions at source. Thus, the policy commitments spelt out in *This Common Inheritance* are given legislative force and include:

- prevent pollution at source;
- minimise risk to human health and the environment;
- encourage the most advanced technical solutions that can be cost effectively applied; and
- use a critical loads approach to pollution in order to protect the most vulnerable sites.

However, there is still some resistance on the part of the UK Government to the extension of pollution control legislation where it involve rigorous examination of the Government's own policies and programs, for example concerning agriculture and energy.

Fourthly, consistent with the Government's privatisation/deregulation ethic, is the view that pollution control legislation should be more flexible. That is, industry should not be straightjacketed by detailed regulations since it is more efficient and more effective to make the market work for the environment. In the words of the Second Year Report on the Government's Environmental Strategy:

> Economic instruments are an inherently more flexible and cost effective way of achieving environmental goals (than is regulation). The Government believes that the time has now come to deploy them more fully to achieve environmental objectives. [...] In future there will be a general presumption in favour of economic instruments (UK Government 1992).

2.2 Legislative Background

UK environmental law has a long history, dating back to the fourteenth century in the case of water pollution control, although it was the public health campaigns of the nineteenth century in response to the Industrial Revolution that shaped environmental protection as we know it today. Legislation has developed progressively since, gaining pace after 1945 and especially from the early 1970s. The various Acts each dealt primarily with a given environmental medium, establishing separate control regimes for each. Examples include, for atmosphere, the 1956 and 1968 Clean Air Acts, and for water, the 1951, 1960, 1961, 1963, 1973 and 1989 Acts.

From its title, the *1974 Control of Pollution Act* would appear to herald an overall approach to pollution control, but that was not so. The Act was little more than consolidating legislation since each environmental medium was dealt with in a separate part (Part I, waste disposal; II, water pollution; III, noise; IV atmospheric pollution; and V, miscellaneous).

Pollution control policy developed rapidly from the mid-1970s, the pace of growth being responsive to and reflective of the strategic significance of the European Community, whose first environmental action plan was agreed in 1973. Whilst the objectives of environmental legislation in the UK are increasingly developed in collaboration with the European Community and organisations such as the United Nations and the Organisation for Economic Co-operation and Development (OECD), the mechanisms through which they are attained are determined nationally. The key piece of legislation is the *1990 Environmental Protection Act*. Part I of the Act defines 'pollution of the environment' as 'due to the release (into any environmental medium) from any process or substances which are capable of causing harm to man or any other living organisms supported by the environment'. This Act, with much of its content motivated by EC activities, drew together and overhauled the regulatory structures and requirements of environmental protection in the UK and introduced *Integrated Pollution Control* (IPC). The objective of IPC is to prevent or minimise pollution in the context of the effect on the environment as a whole, with emphasis on preventive rather than curative measures. IPC, progressively introduced between 1 April 1991 and 1 November 1995, for the most potentially polluting or technically complex industrial processes (Department of the Environment/Welsh Office 1993), therefore sets a new guiding philosophy for UK pollution control.

2.3   The Structure of Responsibilities

Although the Department of the Environment, which exercises a co-ordinating role in respect of pollution control within the wider environmental context, is the UK ministry predominantly for environmental affairs, at the central government level responsibility for action against polluters is shared by several ministries (Figure 2.1). A supporting role is played by a variety of advisory committees, for example the Royal Commission on Environmental Pollution, established in 1970.

In reality IPC is only applicable to prescribed industrial processes and substances, with Her Majesty's Inspectorate of Pollution (HMIP) (1991) the enforcing and authorising authority. Other central government agencies remain responsible for a given environmental medium, examples being the Waste Regulation Authorities for land, and the National Rivers Authority for water. This horizontal division of responsibility is typical of environmental administration. In the case of air pollution, local authority environmental health or pollution control departments deal with domestic and commercial sources as well as those industrial sources which are not the responsibility of the HMIP. In certain circumstances interaction is required with the planning authorities. Thus local authority planning departments are required to consult with the Health & Safety Executive over hazardous industrial installations.

The central government formulates pollution control policy, exercises budgetary control, promotes legislation and advises the various pollution control authorities on policy implementation. Executive responsibility for pollution control is divided between local (municipal) authorities and central government agencies. Before the *1990 Environmental Protection Act,* legislative controls over pollution took the form of a patchwork of controls operated by a patchwork of official organisations, each concerned with a single medium and sometimes with a single type of pollutant, such as radioactive substances or hazardous wastes. This situation changed little after the Act, since pollution controls are still administered by a number of authorities and agencies and through a variety of mechanisms, including licensing and authorisation procedures which target particular processes and substances which can have potentially harmful effects on the environment. Such fragmentation of responsibility poses obvious problems of co-ordination and proposals have been published some time ago (October 1994) in the *Environmental Agencies Bill* for the creation of a new central authority to control pollution. The proposed Environmental Agency will combine all the functions and responsibilities of the National Rivers Authority, Her Majesty's Inspectorate of Pollution, and the Waste Regulation Authorities. This is still some way short of a fully comprehensive pollution control agency.

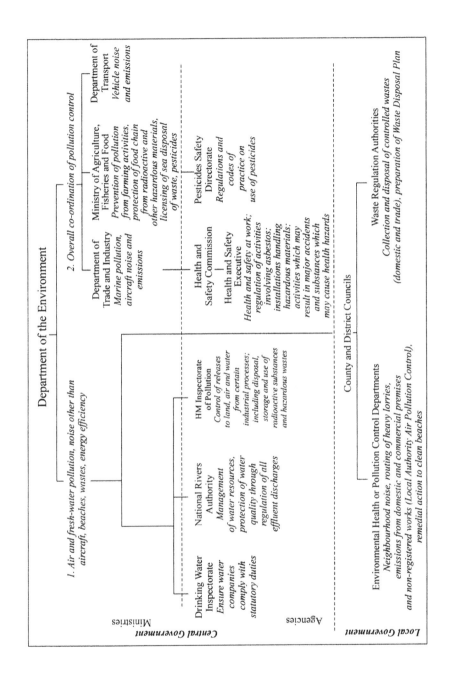

Figure 2.1 Responsibilities for pollution control (England and Wales)

2.4   Principles Underpinning Current Pollution Control

There are three basic approaches to pollution control and abatement:

1   pollution can be prevented by the use of alternative or new, clean 'hub' technologies (or by prohibiting certain activities and use of particular substances);
2   pollution events can be treated using 'end of pipe', clean-up technologies;
3   pollution sources and pollution sensitive receptors can be spatially separated.

In the UK, pollution control policy has been moving towards the first of these, although limits on technological knowledge mean that the other two approaches still have a role to play. Six general principles can be distinguished as underpinning UK pollution abatement policy:

1   sustainability (of environment and resources);
2   the precautionary principle, that is, where there are significant risks of damage to the environment, pollution control will take into account the need to prevent or limit harm, even where scientific knowledge is not conclusive;
3   environmental assessment, involving a systematic evaluation of the potential environmental effects of a project or development and determining the scope for modifying or mitigating those effects to be undertaken before a decision is made to proceed (Department of the Environment 1991);
4   prevention is better than cure;
5   Polluter Pays Principle, meaning that the polluter should bear the expense of carrying out measures decided by public authorities to ensure that the environment is in an acceptable state;
6   no-fault liability, which is operative in waste disposal legislation, whereby the producer of waste which damages the environment is liable for that damage even if not negligent.

These general principles need to be more clearly defined at the implementation stage of pollution abatement policy and three particular concepts dominate UK policy measures since the nineties:

1   Integrated Pollution Control (IPC) came into effect from 1 April 1991 for all new prescribed processes and existing processes undergoing substantial change, with controls over other existing prescribed processes being phased in over a five-year period (Department of the Environment/Welsh Office 1993). A key element of IPC is the Best Practicable Environmental Option (BPEO), which embodies the concept that pollutants must be disposed of so as to minimise effects in all three environmental media, thus achieving the optimal environmental solution overall (Royal Commission on Environmental Pollution 1988).

2   Best Available Techniques Not Entailing Excessive Cost (BATNEEC) is applied to the most harmful pollutants to prevent their release to specified environmental media or, if that is not practicable, to minimise and render harmless such releases. Normally BAT will apply but the presumption can be modified where the costs of applying BAT are excessive in relation to the environmental protection or improvement achieved. BATNEEC has replaced Best Practicable Means (BPM), which had been the fundamental criterion of UK pollution regulation for over a century. Provisions therefore exist to require account to be taken of improved technology and knowledge about the effects of pollution, and IPC authorisations contain conditions requiring the use of BATNEEC, including the achievement of BPEO where releases to more than one environmental medium are likely to occur.

IPC and BATNEEC target the most polluting industrial processes and substances and are administered by the HMIP (1991).

3   Nuisance. In environmental law a 'nuisance' is an act which causes material inconvenience, discomfort or harm and is persistent or likely to re-occur. Statutory nuisances, including odour, smoke, noise, have been identified in various Acts which impose duties on local authorities to detect, investigate and take action. Public nuisances (a tort and a crime punishable by law) and private nuisances (a tort for which remedy is damages) are within Common Law. The distinction between these is largely based on the number of persons affected, with private nuisance usually relating to a single individual's right to use and enjoy their property.

The conceptual basis of pollution policy continues to evolve and further refinements may be anticipated. Two particular developments are likely in the UK. Firstly, the precautionary principle, advocating anticipatory environmental protection, will gain increasing acceptance. It acknowledges that actions for the protection of the environment taken on the basis of our present knowledge may be insufficient and therefore requires that uncertainties beyond our knowledge be taken into account in current pollution policies. That is, even where there is no scientific evidence to prove a causal link between emissions and environmental damage releases should be prevented. Secondly, with improving scientific knowledge refinement of the concept of environmental capacity is to be expected so that maximum allowable concentrations (MAC) or critical loads are more clearly identified and related to environmental quality criteria, including 'acceptable change'.

2.5   Implementing Pollution Control Policy in the UK

There is a choice of measures or instruments through which pollution control policy can be implemented. These may be classified as:

1   regulatory instruments (licensing, permits, setting of standards, bans, zoning, use restrictions, abatement notices);
2   economic instruments (charges, taxes, subsidies, enforcement incentives, deposit-refund schemes, artificial markets);
3   voluntary or suasive instruments (agreements, covenants).

Regulatory instruments are legislative measures aimed at influencing directly the environmental performance of polluters by (a) regulating processes and products used, for example via IPC; by (b) limiting, such as vehicle exhaust emissions, or prohibiting, as with straw and stubble burning after harvest, the discharge of certain pollutants; and/or by (c) restricting activities/use to certain times, such as night flights from airports, or to areas, as in the case of smoke control areas. The main feature of regulating instruments is that a specific level of pollution/ abatement is prescribed and polluters have no choice but to comply or face penalties and fines.

Use of regulatory controls has dominated the approach in the UK, since policy inception until recently, when there has been a swing to favour a market-based approach. The use of economic instruments is consistent with the Government's underlying commitment to the operation of the free market and its belief in its compatibility with environmental excellence (Department of the Environment 1993). The most harmful pollutants must be prohibited by regulatory control but, beyond these, the question becomes how best to deal with the less harmful pollutants. This is where economic instruments now appear to be favoured. Considerable research effort has gone into this (DTI Deregulation Unit 1992; Environmental Resources 1992; London Economics 1992; Coopers & Lybrand 1993).

Implementation distinguishes the most harmful pollution situations (existing and potential) from the less serious, establishing national agencies to 'police' the former and leaving local authorities responsible for the remainder. For example, IPC only applies to those Prescribed Processes and Substances listed in the 1991 Regulations produced under the *1990 Environmental Protection Act*, whilst major industrial hazards are distinguished from notifiable hazards on the basis of the amount of the hazardous material(s) used or stored on industrial premises. Local authorities exercise control of any industrial processes, falling outside IPC, which make emissions to the atmosphere: via *Local Authority Air Pollution Control* (LAAPC) they impose conditions, where appropriate, specifying emission limits, monitoring requirements and operational controls, as well as determining chimney and vent heights to ensure the adequate dispersal of any residual pollutants. Similarly the National Rivers Authority controls effluent discharges from processes outside the HMIP's jurisdiction as well as from water and sewerage companies and all other types of dischargers. An enhanced licensing system for the management of controlled waste (which includes household wastes) came into force on 1 May 1994, implementing the EC Waste Framework Directive. This is operated by the Waste Regulation Authorities and reflects the UK Government's general policy towards waste management based on the reduction, re-use, recovery, safe disposal hierarchy, and subscribes to the *proximity principle* under which waste should be disposed of or otherwise managed as close as possible to the point at which it was generated (Department of the Environment 1994).

The setting of standards is integral to pollution control, especially the regulatory approach. Three types of standard are possible:

1   environmental quality or ambient standard which sets a specified quality level for the environment in a particular locality, that is a permissible level of pollution, which is a legally binding standard not to be exceeded, an example being the maximum concentration of dust per cubic volume of air in a locality;
2   emission or performance standard which sets the amount of pollutant permitted to be discharged from a pollution source, representing the objective to be accomplished, for example limit the amount of dust that can be emitted per unit of time from a given factory;
3   specification or design standard which sets a requirement to comply with a design specification, or how an objective must be accomplished, for instance require installation of the best available dust-removing equipment.

Both environmental quality and performance standards leave the polluter to decide how to conform: the UK favours the use of performance standards whereas the EC prefers environmental quality standards. Most UK emission standards were *presumptive limits* set after consultation with the industry concerned and, because not written into the legislation, could be easily updated. *Statutory limits* embodied in legislation, following EC directives, may well be less flexible.

2.6   Pollution Control and Land-Use Planning

The British land-use planning system complements pollution control policy through its role in determining the location of potentially polluting developments. The potential for pollution from a development to affect adversely the environment and the use of adjacent sites for other development can be a material consideration in deciding whether to grant planning permission (Department of the Environment 1994). The role of the planning system therefore focuses on whether the potentially polluting development is an acceptable use of land rather than on the control of the processes or substances involved. In deciding whether to grant planning permission the local planning authority must satisfy itself on these land-use grounds, safe in the knowledge that concerns about potential releases of pollutants can be left for the pollution control authorities to take into account when considering the application for authorisation or licence.

Once a potentially polluting development has been allowed to proceed, the local planning authority has no control over further improvement of its environmental performance, but the authority will be concerned with decisions about what other development might be permitted in proximity to the potential source of pollution. Likewise for any existing polluters where any tightening of environmental performance measures is dependent upon the exercise of powers by the pollution control authorities. Local planning authorities do, however, have a role to play in regulating what happens after any potentially polluting development ceases to use a site, seeking to ensure that land and water resources are restored to a condition rendering them capable of an agreed after use. This latter approach is consistent with the government's policy towards tackling contaminated land, namely, a 'suitable for use' approach (Department of the Environment 1994).

2.7   Future Directions for UK Pollution Policy

The current state of UK pollution policy suggests a more holistic approach is being adopted, as well as the greater reliance on market-based instruments. Already in the case of IPC, polluters gain the important practical benefit of 'one stop shopping', having to deal with only one enforcing authority. This is the HMIP, which consults other national agencies and imposes conditions that they require. This model as it was introduced is only applicable to the most harmful pollution situations. The establishment of the proposed Environment Agency is a step in the right direction and should encourage wider dissemination of good practice, but questions about the adequacy of its powers will only be tested once it comes into being. Furthermore, legislation has limited influence unless effectively enforced and monitored and there is evidence that insufficient resource is devoted to pollution control: for example, the HMIP is understaffed (National Audit Office 1991). The new Environment Agency may not fare any better.

In the light of this, the increased emphasis on a market-based approach offers flexibility, placing greater onus on the polluter to adjust. Certainly, if economic instruments are sensitively tuned, cost-effective and innovative improvements in combating pollution will result. Indeed, increased use of voluntary instruments and environmental evaluatory techniques is to be expected as companies adopt 'green' business practices. So far the UK Government has adopted a cautious approach to the introduction of economic instruments into pollution control policy compared to other countries, as OECD surveys reveal (Opschoor 1994). The basic command-and-control approach of regulation will remain a cornerstone of British pollution control policy for the foreseeable future despite the recognition given to the value of a market-based approach.

Finally, greater harmonisation of coverage of UK and EC pollution policy will occur (Haigh 1992), although the principle of subsidiarity and powers of derogation will ensure that certain differences in implementation measures and timing of actions will remain.

Note

1   Brian Goodall is Professor and Head of the Department of Geography; Director of the Tourism Research & Policy Unit, and Consultant Director of the NERC Unit for Thematic Information Systems at the University of Reading, England.

References

Central Office of Information (1993) *Pollution Control*, HMSO, London.

Coopers & Lybrand (1993) *Landfill Costs and Prices: Correcting possible market distortions*, HMSO, London.

Department of the Environment (1991) *Policy Appraisal and the Environment*, HMSO, London.

Department of the Environment (1993) *Making Markets Work for the Environment*, HMSO, London.

Department of the Environment (1994) *PPG 23 Planning and Pollution Control*, Department of the Environment, London.

Department of the Environment/Welsh Office (1993*) Integrated Pollution Control: A practical guide*, HMSO, London.

DTI Deregulation Unit (1992) *Checking the Cost to Business: A guide to compliance cost assessment*, Department of Trade and Industry, London.

Environmental Resources Ltd. (1992) *Economic Instruments and Recovery of Resources from Waste*, HMSO, London.

Haigh, N. (1992) *Manual of Environmental Policy: The EC and Britain*, Longman, Harlow.

Her Majesty's Inspectorate of Pollution (1991) *Chief Inspector's Guidance to Inspectors: Environmental Protection Act, 1990: Industrial Sector Guidance*, Department of the Environment, London. (IPR 1: Fuel and power industry sector; IPR 1/1 Combustion processes; IPR 2: Metal industry sector; IPR 3: Mineral industry sector; IPR 4: Chemical industry sector; IPR 5: Waste disposal industry sector.)

London Economics (1992) *The Potential Role of Market Mechanisms in the Control of Acid Rain*, HMSO, London.

National Audit Office (1991) *Control and Monitoring of Pollution: Review of the pollution inspectorate*, HMSO, London.

Opschoor, J.B., A.F. de Savornin Lohman and H.B. Vos (1994) *Managing the Environment: The role of economic instruments*, Organisation for Economic Co-operation and Development, Paris.

Royal Commission on Environmental Pollution (1988) *Best Practicable Environmental Option*, Cm 310, HMSO, London.

U.K. Government (1990) *This Common Inheritance*, Cm 1200, HMSO, London.

U.K. Government (1992) *This Common Inheritance: Second year report*, Cm 2068, HMSO, London.

U.K. Government (1994a) Sustainable Development: The U.K. Strategy, Cm 2426, HMSO, London.

U.K. Government (1994b) *Climate Change: The U.K. programme*, Cm 2427, HMSO, London.

U.K. Government (1994c) *Biodiversity: The U.K. action plan*, Cm 2428, HMSO, London.

U.K. Government (1994d) *Sustainable Forestry: The U.K. programme*, Cm 2429, HMSO, London.

World Commission on Environment and Development (1987) *Our Common Future*, Oxford University Press, Oxford.

Chapter 3

Tackling the Problem of Conflicting Land Uses in Hong Kong: A Planner's View

Pun Chung Chan[1]

Summary

At the beginning of the nineties, continuous efforts have been made to resolve the problem of conflicting land uses in the old urban areas in Hong Kong through the plan making and development control processes. Existing mechanisms under the town planning and land legislation are not entirely effective. The application of existing mechanisms and their limitations are briefly explained.

In order to overcome the existing problems, new efforts are being identified and investigated to evaluate their effectiveness and practicability in application. Some of these efforts would require policy and legislative changes before they can come into effect. A case study has been chosen to illustrate the existing problems and how some of the problems have been alleviated by the introduction of a new concept of Dual-Purpose Industrial-Office Buildings permitted in industrial zones.

3.1   Introduction

In Hong Kong, there are many problems of conflicting land uses in the old urban areas. Examples are industrial buildings in juxtaposition with residential buildings, potential hazardous installations such as oil depots, cement works next to a housing estate, and vehicle repair workshops within residential buildings. These uses are non-conforming uses, but many of them are however legal uses under the statutory town planning system in Hong Kong, if they were already in existence before the publication of statutory town plans. There are limited powers available in the planning legislation to control or eliminate such uses. Existing non-conforming uses are allowed to continue to exist. They are only required to conform with the statutory town plan upon a material change of use or redevelopment of the land or building in question.

During the nineties, continuous efforts have been made to resolve the problem of conflicts between land uses in both the plan making and development control processes. Further efforts are being made to identify new mechanisms and to introduce changes to the legislative framework, where necessary, to tackle the problem.

3.2   Planning Control

Statutory planning control in Hong Kong is governed by the Town Planning
Ordinance. The Ordinance makes provision for the appointment of a Town Planning
Board (TPB) to prepare statutory town plans (Town Planning Office 1989). Definitive
land use zones are shown on the town plans and the notes attached to each plan
specify, for each zone, the uses which are always permitted and uses which may be
permitted by the TPB, with or without conditions, upon application. If a proposed
development is 'a use always permitted' in the land use zone (i.e. Column One on the
notes), there is no need to apply for planning permission from the TPB and the
development can proceed as long as it complies with the Buildings Ordinance, the
lease and requirements under other relevant legislation. If the proposed development is
'a use which may be permitted' under the town plan upon application (i.e. Column
Two use on the notes), planning approval from the TPB must be obtained. Otherwise,
the building plans of the proposed development would be rejected under the Buildings
Ordinance (Town Planning Board 1993). If the proposed development is a use not
mentioned in the notes attached to the statutory town plan, then the development
cannot proceed.

Planning control for new development and redevelopment under the planning
legislation relies upon the Buildings Ordinance's power to reject the building plans
which are not in conformity with the statutory town plan or planning permission
granted by the TPB. There is however limited power in both the Town Planning
Ordinance and the Buildings Ordinance to deal with existing uses.

3.3   Existing Mechanisms

Under the provisions of the Town Planning Ordinance, there are certain mechanisms
which have been used to resolve the problem of conflicting land uses (Planning,
Environment and Lands Branch 1991). These include:

Application for Change of Use

There are provisions in the notes attached to statutory town plans for applications for
change of use. Factory buildings close to residential buildings may be permitted to
redevelop into commercial/office buildings with planning approval from the TPB.
Obtaining such approval and implementation of the redevelopment proposal are
however decisions of the landowners.

Up-zoning

Land may be up-zoned to a higher value use on statutory town plans e.g. from industrial to commercial or residential uses, so as to provide a greater incentive for early termination of an environmentally intrusive use. Even with such incentive, the timing of redevelopment and thus termination of the use cannot be guaranteed. Planning incentives alone cannot solve the problem particularly when buildings in fragmented ownership are involved.

Comprehensive Redevelopment

In some cases, especially when redevelopment of non-conforming buildings is hindered by fragmented ownership and difficulties in site assembly, removal of such buildings may be brought about through larger scale comprehensive redevelopment by quasi-government urban redevelopment corporations. Such sites will be zoned as Comprehensive Development Area on the statutory town plan. Redevelopment in a comprehensive manner may be permitted by the TPB which also has the power to recommend compulsory purchase of properties within the Comprehensive Development Area.

Compulsory Purchase

In some extreme cases, it may be justifiable to resume land taken up by non-conforming uses or buildings that are causing serious environmental problems. To do so, the land has to be zoned for a public purpose on the statutory town plan so that the TPB can make recommendations to resume land under the land acquisition legislation.

3.4   New Efforts

The problem of conflicting land uses cannot always be solved by the above mechanisms which have their own limitations. Additional methods and new approaches are being investigated.

Dual-Purpose Industrial-Office (I/O) Building

A new concept of dual-purpose I/O building was recently introduced. Suitable industrial sites may be allowed under the planning permission system to redevelop into a dual-purpose I/O type of building (Town Planning Board 1994). The building is designed to be used flexibly for both industrial and office purposes. It is intended that such buildings would accommodate non-polluting types of clean industries, quasi-industrial operations and ancillary office activities related to these industrial/quasi-industrial operations. This concept represents an effort to recognise the changing industrial structure in Hong Kong as a result of close co-operation with industrial activities in South China and an attempt to encourage the construction of more environmentally acceptable buildings at sensitive locations. A set of TPB's guidelines for application is at Annex A.

Industrial (Environmentally Sensitive) (I/E) Land Use Zone

A preliminary study has been undertaken to look into the feasibility of introducing a new I/E zone for industrial sites locating in close proximity to environmentally sensitive uses. The list of industrial uses permitted in the I/E zone is intended to be very restrictive to include only non-polluting clean industries. Marginally clean industries would require planning approval from the TPB. The concept is likely to be met with opposition from industrial operators and landowners because the zone restricts the range of permissible industrial uses. There are also problems of defining and updating the permitted uses in the notes for this zone. Further research on how the planning, land and building authorities can devise a workable system to implement the concept is the next step to be carried out.

Designation of Environmentally Sensitive Land Use Area

Investigation is being carried out to consider the feasibility of designating land use area which is environmentally sensitive to surrounding development on the statutory town plan as a special area so that all new development or material change of use within the area would require planning permission (Planning, Environment and Lands Branch 1991). Such applications for permission should also be supported by relevant environmental assessment reports.

Amortisation

A study is being undertaken to examine the practicability of introducing the concept of amortisation in Hong Kong (Planning Department 1993). Amortisation in this context refers to the compulsory termination without compensation of a non-conforming use at the end of a specified period of time. The basic idea is to allow the operator of a non-conforming use a specified grace period to continue and amortise the investment, after which the non-conforming use must be discontinued or changed to conform to the statutory town plan. Introduction of the amortisation concept will require amendments to the planning legislation.

3.5   Case Study 'Ma Tau District'

In the Ma Tau Kok district which is an old urban area situated right at the heart of the city (Figure 3.1), there are many examples of conflicting land uses. These include factories in juxtaposition with residential buildings, conversion of retail shops on the ground floor of residential buildings into workshops and vehicle repair garages, as well as an old gas plant being located in close proximity to residential buildings. The existing mechanisms that are available to the planning and land authorities outlined in the foregoing paragraphs have been applied to encourage change of use and redevelopment in accordance with the planning. This redevelopment includes changes from non-conforming uses to conforming ones, redevelopment of factories to commercial/residential buildings and comprehensive redevelopment of amalgamated sites to achieve environmental improvement through careful planning and building design.

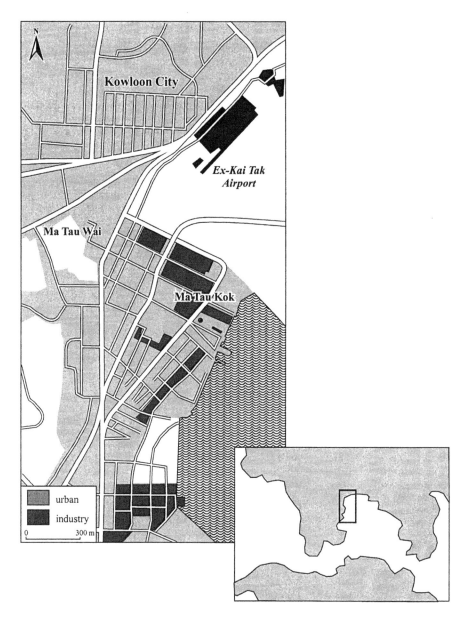

Figure 3.1   Town planning ordinance, Hong Kong Town Planning Board, Kowloon planning area, Ma Tau Kok

Many of the new efforts are at the various stages of study and they cannot readily apply to the Ma Tau Kok district. However, the concept of Dual-Purpose I/O Building has been introduced by the TPB for some years. Some clusters of industrial buildings in the district (Figure 3.1) adjacent to residential buildings are being redeveloped into I/O Buildings thus reducing their impacts on the adjacent development. Such redevelopment is taking place here because the district, being close to the airport, is suitably located to accommodate clean types of industries, warehouses related to air freight activities and office activities ancillary to industrial operations. It is expected that this trend of redevelopment would continue.

3.6   Conclusion

The problem of conflicting land uses in existing urban areas cannot be solved overnight. It requires co-ordinated efforts from the planning, environmental and land authorities to provide the institutional framework, enforcement mechanisms and suitable incentives to achieve the required changes. The prevailing economic, social and political aspirations are also important considerations and they can affect the pace of change. Opportunities have been taken of the growing economy in Hong Kong to encourage change of use and redevelopment which have resulted in certain improvements. However, further work has to be carried out and new approaches and mechanisms have to be established to tackle the more problematic uses.

Note

1   Pun Chung Chan is Government Town Planner at the Planning Department in Hong
         Kong.

References

Government Secretariat Planning, Environment and Lands Branch (1991) *Comprehensive Review of the Town Planning Ordinance - Consultation Document*, Government Secretariat Planning, Hong Kong.

Planning Department (1993) *Study Brief on Amortization of Non-Conforming Existing Uses*, Planning Department, Hong Kong.

Town Planning Board Hong Kong (1993) *Town Planning Board Annual Report 1992*, Town Planning Board Hong Kong, Hong Kong.

Town Planning Hong Kong (1993) Kowloon Planning Area No. 10 Ma Tau Kok Outline Zoning Plan No. S/K10/7 and Explanatory Statement, Town Planning Hong Kong, Hong Kong.

Town Planning Board Hong Kong (1994) Town Planning Board Guidelines for Application for Composite Industrial-Office Buildings in Industrial Zone under Section 16 of the Town Planning Ordinance, Town Planning Board Hong Kong, Hong Kong.

Town Planning Office Hong Kong (1989) *Town Planning in Hong Kong*, Town Planning Office Hong Kong, Hong Kong.

[Annex 3.1]

TPB PG-NO. 4 (revised Version 1/94)

Town Planning Board Guidelines for Application for Composite Industrial-Office Buildings in Industrial Zone under Section 16 of the Town Planning Ordinance

Important note

The guidelines are intended for general reference only. The decision to approve or reject an application rests entirely with the Town Planning Board and will be based on individual merits and other specific considerations of each case.

Any enquiry on this pamphlet should be directed to the Planning Information and Technical Administration Unit of the Planning Department, 16/F, Murray Building, Garden Road, Hong Kong - Tel. No. 8482402.

These guidelines are liable to revision without prior notice. The Town Planning Board will only make reference to the guidelines current at the date on which it considers an application.

1 Scope and Application

1.1 The Town Planning Board defines composite industrial-office (I/O) building as a dual-purpose building in which every unit of the building can be used flexibly for both industrial and office purposes. In terms of building construction, the building must comply with *all* the relevant building regulations applicable to both industrial and office buildings, including floor loading, compartmentation, lighting, ventilation, provision of means of escape and sanitary fitments.

1.2 The Town Planning Board recognises the need to meet the demands of an industrial sector which is undergoing structural change. It has become evident in recent years that there has been a trend moving towards high-technology and high-quality products, and the relocation of part of the labour-intensive production process to areas such as China and other South-east Asian countries. Such changes have resulted in an increased demand amongst industrial firms for more floorspace to be used for management, administration, design, research and quality control, in addition to storage and other industrial activities which cannot be accommodated in normal commercial buildings. It is envisaged that such trend of changes would continue and this creates the need for providing accommodation which can be used flexibly for both industrial and office purposes.

2 Uses Permitted in the Composite Industrial-Office Building

(a) Industrial operations, quasi-industrial operations and ancillary office activities related to these industrial/quasi-industrial operations will be permitted in the composite building. Examples of such uses are industries requiring extensive design, research, development and testing, quality control and data processing inputs; industries whose labour-intensive production activities have been relocated outside Hong Kong, retain locally only such activities as research and design, quality control, sample making, parts assembly, sales and distribution; industries of a higher level of technology requiring a better working environment; firms providing technical services to industrial operations requiring floor space to be set aside for such activities as workshops and laboratories.

(b)   Heavy and noxious industrial operations will not be permitted, in order to maintain a reasonably clean and comfortable internal environment for the composite building. Pure office activities unrelated to any industrial operations will not be permitted.

(c)   As industrial operation should still remain as the predominant use in the composite building, the general intention is to discourage the infiltration of general commercial uses as restaurants and supermarkets. From the fire safety point of view, it is necessary to avoid unnecessary agglomeration of members of the public who are not familiar with the particular building and the industrial neighbourhood. Apart from offices related to industrial and quasi-industrial operations, all other commercial uses are subject to planning permission from the Town Planning Board under Section 16 of the Town Planning Ordinance and each case will be considered on its individual merits. Where considered appropriate, a limited provision of floorspace for such necessary and complementary local services to the industrial area as banks, showrooms and local provisions stores may be allowed by the Town Planning Board upon application.

3   Size and Location of Ancillary Offices

Under the I/O concept, there is no restriction on the size of an office to be established within an I/O building as long as it is ancillary and direct related to an industrial operation.

4   Main Planning Criteria

(a)   Every unit within the composite industrial-office building should be designed, constructed and made suitable for both industrial and office uses; that is to say, a pure office building or a pure industrial or a building with discrete horizontal and/or vertical segregation into purely office and industrial portions will not be allowed. The general principle is that where building design requirements, e.g. provision of prescribed windows, sanitary fitment, lighting, ventilation and fire resistance etc., for industrial and office buildings differ, the more stringent

requirements must be adopted. Small sites of elongated shape with single and narrow frontage which would be unlikely to satisfy the design requirements are generally not supported.
(b)   Separate entrances and lift lobbies for goods and passengers must be provided.
(c)   Car and goods vehicle parking, loading/unloading requirements must be provided in accordance with the Hong Kong Planning Standards and Guidelines as if the whole building were an industrial building. For applications requiring the conversion of existing industrial buildings or redevelopment of industrial sites which are severely constrained by site characteristics or configuration, applicants should consult Commissioner for Transport to explore a practicable and suitable level of industrial parking requirements prior to the submission of the planning applications.
(d)   Spaces for the parking of private cars for office users must also be provided in accordance with the Hong Kong Planning Standards and Guidelines and to the satisfaction of Transport Department. However, special consideration may be given to relax the parking requirements for office users if the size, dimension and configuration of the application site are so constrained that it is difficult to comply with the standard requirements. For such applications, applicants should consult Commissioner for Transport to work out a practicable and suitable level of parking provision prior to the submission of the planning applications.
(e)   The application sites must be easily accessible to public transport.

5   Interim Arrangement prior to the Availability of Composite Industrial-Office Buildings

Since purpose-designed industrial-office development is only a concept recently approved, and it will take some time before industrial-office buildings are completed for occupation, the Town Planning Board may give favourable considerations in the interim to applications for industrial uses requiring large ancillary office spaces to be temporarily accommodated in industrial buildings, pending the availability of suitable floorspaces within industrial-office buildings.

6   In-Situ Conversion of Industrial Building to Industrial-Office Use

The industrial-office building development application is not restricted to redevelopment of the whole site. In-situ conversion of existing building to I/O use will also be considered by the Board provided that the conversion satisfies the main planning criteria in para. 4.

Town planning board
January 1994

Chapter 4

From Measurement to Measures: Land Use and Environmental Protection in Brooklyn, New York

N. Anderson, E. Hanhardt and I. Pasher[1]

4.1   Introduction

Walk along the East River's edge in Brooklyn, New York's Greenpoint/Williamsburg district, and Manhattan's picture perfect skyline seems just out of reach. Turn around, and you could be in another country. Here, Spanish, Polish and Yiddish are commonly spoken and store signs are often in a language other than English. Greenpoint/Williamsburg is a working class neighbourhood, with large pockets of poverty; home to many recent immigrants. According to the 1990 census, the district has a residential population of 155,000 and an average annual household income of $18,900. This area, designated Brooklyn Community Board #1 by City government, is unified in name, but is not often unified across neighbourhood or ethnic lines. However, enforced coexistence with an exceptional number of factories and municipal facilities, and concerns over industrial pollution, have combined to coalesce the community and create a groundswell of demand for change in the way government plans how urban districts are developed and how environmental protection is realised.

Greenpoint/Williamsburg is a microcosm of the problems facing cities all around the world – each of which needs to promote sustainable development, protect and preserve its diverse neighbourhoods and foster social and environmental equity. We believe that these problems have emerged from the disjunction between rules and procedures governing land use and those governing environmental protection within an urban political economy where wealth and influence are strongly tied to the dynamics of the real estate market.

4.2   The History and Character of Greenpoint/Williamsburg

Greenpoint/Williamsburg developed much of its current character in the nineteenth century as a centre of shipbuilding, oil refining, manufacturing, storage and transport. As was common in the nineteenth century city, factories, oil depots and other manufacturing related establishments were clustered together with housing. People lived where they worked. Greenpoint/Williamsburg's character did not change over the years. To this day this area has the highest 'walk-to-work' ratio in New York City. The mixed use character of Greenpoint/Williamsburg was incorporated into the Zoning Resolution in 1973.

Today, as a centre of urban manufacturing and allied industries, Greenpoint/Williamsburg is home to many firms, large and small, that use a wide array of hazardous substances. Typical industries include metal platers, woodworkers and manufacturers of chemical dyes and plastic consumer goods. Greenpoint/Williamsburg also houses the only operating facility in New York City permitted to store low level radioactive and hazardous waste, separated from each other by only a steel door. This facility also shares a common wall with a residential structure and is located down the block from a public school and a playground. The construction of the public school predated the opening of this facility; however, the facility itself predated present environmental laws, and was grandfathered when the new regulation was adopted. Some residents would like to protect the local environment by having all manufacturing move out regardless of the impact on jobs and the local economic base. The majority of residents who also depend on those manufacturing jobs for their livelihoods, would prefer to find mutually agreeable solutions to the problems.

Compounding the environmental impact of being the location of the greatest concentration of manufacturers possessing hazardous materials in New York City is the presence here of the largest sewage treatment plant on the Eastern seaboard and the City's last remaining municipal incinerator including a marine transfer station. Also burdening this district, which contains only 3 per cent of the City's land, is a 17 million gallon underground oil plume, 28 per cent of the commercial solid waste transfer stations, the only liquid natural gas storage facility, the largest number of oil storage tanks in the City, and the proposed site of what will be, if built, the largest incinerator on the Eastern seaboard.

4.3   Land Use and Environmental Protection

The modern environmental protection movement is now a few decades old. Its statutes and regulations set goals for environmental quality as well as for the protection of human health. These goals are met by control of specific pollutants. Such control has been achieved either by banning certain substances or by setting ceilings on permissible emissions and concentrations of designated pollutants.

Zoning laws in New York City which govern land usage originated in 1916. The zoning rules which govern land use today were adopted in 1961, and prior to passage of modern, scientifically informed environmental protection laws. In New York City, while the Department of City Planning promulgates the zoning rules and conducts hearings for approval of certain proposed projects and Zoning Map changes, actual conformance to zoning regulations is enforced by the Buildings Department. Although the Department of Environmental Protection's (NYCDEP) regulations are referenced in certain of the Zoning Resolution's performance standards, there is little if any co-ordinated enforcement of these standards between the Buildings Department and NYCDEP. The result of this legislative and regulatory disparity is the concentration of environmental and quality of life threats that characterise neighbourhoods like Greenpoint/Williamsburg and, of equal importance, make the solution to these problems hard to achieve.

Zoning rules restrict the siting of most environmentally intrusive facilities to 'M' zones. 'M' zones are parcels of land where industrial uses including manufacturing and municipal public works such as sewage treatment plants may be sited. There are currently three classes of manufacturing zones in New York City: 'M1', light manufacturing, high performance; 'M2', medium manufacturing, medium performance; and 'M3', heavy manufacturing, low performance. These distinctions refer to the stringency of performance standards with 'M3' permitting the most noxious uses. Theoretically, the City can re-designate an area as an 'M' zone or use an emergency zoning override to site a municipal facility, but as a practical matter it does not.

In 1973, a 'Special Mixed Use District' was established which legitimised the development of both residential and industrial uses on the same block in certain areas of Greenpoint/Williamsburg. This designation allowed for moderate density residential uses and manufacturing uses that meet 'M1' performance standards. This zoning has proved to be quite burdensome because its success depends upon the vigorous enforcement of 'M1' standards and environmental laws.

The only existing regulations that attempt to introduce environmental considerations into the land use process are New York's City Environmental Quality Review and State Environmental Quality Review. These city and state laws require projects subject to certain discretionary actions to undergo an environmental review. This review usually takes the form of an Environmental Impact Statement (EIS) that is specific to the project under consideration.

The New York City Charter Revision of 1989 intended to balance the burdens imposed on communities by the siting of noxious or undesirable municipal uses, whereby no single community is forced to absorb a disproportionate share of these uses. Called 'Fair Share', this legal principle is the only local law which addresses 'environmental equity' concerns. Although 'Fair Share' is a good first step, given existing land use and zoning patterns, achieving a 'Fair Share' distribution of environmentally burdensome facilities is highly problematic. Adequate legal and political tools to preserve the host neighbourhood and to improve its environmental quality in a systematic fashion are being developed (Blanco 1998).

Greenpoint/Williamsburg, because of its industrial history and the continued presence of a manufacturing base, has remained predominantly zoned 'M', despite a Citywide trend to rezone land from 'M' to non 'M' classifications due to a general decline in manufacturing. This 'M3' zone concentration has proved useful to city planners and public works builders because it has allowed them to continue to site these facilities in the area without forcing them to deal with the dwindling 'M3' zones in the rest of the City. The implicit theory here is that proposals for municipal public works facilities can be appropriately accommodated in areas zoned to allow such uses. This negates the necessity to employ functional mechanisms to determine the most appropriate site, such as the 'Fair Share' or EIS criteria.

4.4   The Greenpoint/Williamsburg Environmental Benefits Program

In the case of Greenpoint/Williamsburg, a first-of-its-kind mechanism for fact-gathering and remediation to foster environmental equity, the Greenpoint/Williamsburg Environmental Benefits Program (GW/EBP) was established. Funded with $850,000 set aside by New York City as a result of a court settlement for violations of the Newtown Creek sewage treatment plant permit, the New York City Department of Environmental Protection has been working closely with Greenpoint/Williamsburg residents and businesses to design and implement environmental improvement projects that will reduce existing hazards and promote environmentally sound development in the future. Although made up of many particulars, the GW/EBP has four central elements. The first is a two-part epidemiological study of mortality and morbidity in the district. The study helps to determine whether residents of Greenpoint/Williamsburg exhibit a statistically elevated incidence of certain cancers and childhood leukaemias, birth defects and asthma. Study results indicated that certain stomach cancers were found at a rate higher than the City norm.

The second major component of the GW/EBP is a pilot multi-media compliance, enforcement and pollution prevention project. Sixteen facilities have been inspected as a result of the project and three inspections per month are planned. This component of the GW/EBP is the only one that makes NYCDEP take a hard look at its own operations and experiment with new ways to protect the environment using existing resources.

Other GW/EBP pilot projects include the Clean Industry Program, household hazardous waste reduction, and environmental education and outreach are being developed with this fund and other monies obtained from the United States Environmental Protection Agency.

The linchpin and fourth element of the GW/EBP is the development of a state-of-the-art, locally based Geographic Information System (GIS) designed and now being implemented with data to document existing environmental and health conditions in the area. The GIS generates maps which offer a vivid visual display of the concentration and distribution of environmental, health, land use and demographic data. These maps will be used to demonstrate the environmental impacts on the community, to plan for the district's growth, and future enforcement actions.

Perhaps most significantly, in documenting existing environmental risks and burdens, the GIS has substantiated many of the community's concerns over the extent of the environmental problems facing Greenpoint/Williamsburg. As with many urban districts, Greenpoint/Williamsburg is concerned with the problems resulting from the location of multiple pollution sources in a densely populated area. While an individual source may be found to be only a small contributor to air or water pollution, the community is anxious about possible cumulative impacts from many sources.

But there still is no mechanism for assessing and responding to cumulative or aggregate impacts. For example, Environmental Impact Statements (EIS) which are intended to disclose the impacts of proposed projects on the surrounding area, presuppose the existence of a baseline environment assessment as an analytic starting point. However, no such baseline exists for Greenpoint/Williamsburg or for any other portion of the City.

Although an aggregate impact discussion is required as part of a 'Fair Share' analysis, boundaries of the impact area to be studied are again set arbitrarily relative to the size or operational characteristics of the proposed facility. With both 'Fair Share' and EIS it appears that the goal is disclosure, not assessment.

The development of a methodology for conducting baseline environmental assessments will be of benefit to not only Greenpoint/Williamsburg but to all New York City communities suffering from environmental degradation directly or indirectly. After all, while the smell of a chemical pollutant may only reach a block or two, the health impact of that pollutant may extend much further.

4.5   Greenpoint/Williamsburg Aggregate Baseline Environmental Assessment

One of the most important outputs of the GW/EBP will be the study of the accumulated environmental conditions in the district. The GW/EBP will hire a consultant to design and perform this unique aggregate baseline environmental assessment (ABEA). The scope of work calls for four tasks. First, a methodology will be developed for analysing the aggregate impacts which have already been identified and mapped in the GIS. Second, based on but not limited to GIS data, the following data fields will be gathered: environmental quality; census; public health; land use; environmental compliance and enforcement as well as government resource allocation. Third, using the methodology developed in task 1 and the data gathered in task 2, the consultant will analyse the data and produce an aggregate baseline environmental assessment for Greenpoint/Williamsburg. Fourth, the consultant will make specific recommendations for implementing actions in that district. A menu of options will be presented where, in New York City's land use and environmental decision making processes, the ABEA could be incorporated. These options may include zoning, city and state environmental quality regulations, 'Fair Share' criteria, planning documents, capital and expense budget resource allocations, as well as enhanced environmental standards and rules.

Importantly, *for the first time the consequences of the disjunction between existing land use and environmental protection provisions will be examined.* Yet, if the intent is to ultimately close the rift between land use planning and environmental protection, documentation alone will not suffice. The goal of the aggregate baseline environmental assessment must be to go beyond measurement and move into taking legal and political action. The findings and recommendations of the study will have to become an instrument that citizens of Greenpoint/Williamsburg and government decision-makers use in the political arena of land use regulation and planning, economic development planning, and enhanced environmental protection initiatives. If successful, the GW/EBP (this partnership of informed citizens and local government) will have provided an important model to all urban areas for designing and carrying out effective abatement and prevention strategies for aggregate, multi-source and multi-media environmental problems.

Note

1   Nancy Anderson Ph.D. and Eva Hanhardt are working for the New York City Department of Environmental Protection, Community Environmental Development Group. Inez Pasher works for the Community District #1, Brooklyn, New York.

Reference

Blanco, H. (2004) Lessons from an Adaptation of the Dutch Model for Integrated Environmental Zoning (IEZ) in Brooklyn, NYC in D. Miller and G. de Roo (eds.) *Integrating City Planning and Environmental Improvement*, Ashgate, Aldershot, UK.

Chapter 5

Land Use Conflicts and Problem Solving Strategies: Three Case Studies in the Metropolitan Area of Lisbon

J. Farinha,[1] L. Vasconcelos[2] and A. Perestrelo[3]

Summary

The main objective of this chapter is to analyse the current practice in land use planning in dealing with environmental problems caused by intrusive activities. It aims to obtain an overview, identify lessons and assess the strategies that municipalities use in solving environmental conflicts. The chapter focuses on three case studies (see Figure 5.1) specially selected in the Metropolitan Area of Lisbon (MAL): (a) The cement plant of Alhandra located in the middle of a large residential area (VF de Xira) with evident conflicts caused by air pollution, traffic disruption and other noxious spillovers affecting thousands of inhabitants; (b) A bulk of nineteenth-century industries presently located in residential area in Lisbon, and close to the waterfront next to the Tagus river with considerable high amenity potential, producing pollution, spatial disruption and visual intrusion, and presenting risks for the surrounding city areas; (c) The International Exposition 'Expo 98', which is located in a physically and socially declining large area with high risk industries, a garbage treatment plant, run down warehouses and old harbour precinct in the eastern part of Lisbon. The conclusions from these case studies will suggest future research areas and will provide updated information on the potential for application of environmental zoning in the context of MAL.

5.1   Introduction

Chaotic land use is considered one of the most serious threat to the environment in Portugal (Pimenta 1993), while industrial pollution has been identified as the one with more serious environmental impacts (Marques 1992). It is therefore a priority to implement policies to address this problem.

The numerous negative effects produced by the industry include noise, water pollution, soil pollution, air pollution, smell, fire risk and explosion, and potential dangers for safety, health and environment in general. When industries are surrounded by a residential area or other sensitive land uses, the situation becomes even more serious.

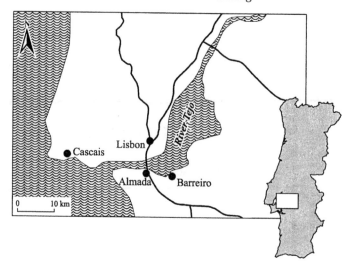

Figure 5.1   Lisbon, Portugal; including the three case study areas

The rapid growth of the Metropolitan Area of Lisbon and a relatively weak control of land use transformation until the late seventies have led to an undesirable mixture of uses. This creates serious conflicts particularly where intrusive activities such as polluting industrial plants are located in areas with environmentally sensitive land uses, such as residential areas. In other cases it was the sensitive land use that organically grew around the aggressive activity. This frequently happened with workers' residential areas embracing their working place.

In Portugal, municipal planning has undergone considerable change due in part to two components: new enacted legislation forcing municipalities to have a municipal land use plan, and the availability of funds resulting from Portugal's entrance in the European Union. The municipalities also have at their disposal new and stronger means for plan implementation.

Therefore, the right setting to intervene in such cases of conflict of uses is created. How did the municipalities deal with these cases? Planning offers an opportunity to solve these type of conflicts of use, while restructuring the urban areas. How was the local planning process developed? What specific efforts emerged to deal with these issues? Did municipalities develop ways to justify these interventions and assure their implementation?

5.2   Portuguese Land Use Planning Instruments at the Municipal Level

The Municipal Land Use Plan (*Plano Director Municipal*) (PDM) specifies the major characteristics of land use throughout the municipality. It works at 1/25,000 or 1/10,000 scales and designates a dominant land use to each parcel of land, making a kind of a flexible zoning. Other activities that are compatible with the dominant land use can be authorised to occur.

For the urban areas of the municipality the PDM is complemented by an Urban Plan and a Detailed Plan. The working scales are between 1/5,000 and 1/2,000. At these scales they define and attribute in detail to each parcel of land an urban use. This use can be the dominant one, defined at the PDM level, or a compatible one.

There exists a large set of sectoral legislation that not only has to be considered by the land use plans, but also contains direct implications for the use of the land and the compatibility among activities. For instance, legislation for industries clearly divides them in four different classes according to their probability of pollution and environmental impact and specifically states that classes A and B, the most polluting, have to be located in industrial zones outside residential areas. The industrial activities from class C are allowed inside residential areas, but located away from residential buildings, while industries from class D can be placed within a residential building.

Another example of sectoral legislation with direct impact on land use regards air quality. There are limits to the amount of air pollution that can occur in one area (immissions control) and also a limit to the amount of air pollution that can be sent into the atmosphere by a certain source (emissions control). Neither of these two limits should be exceeded. The authorisation and location process of a certain activity does not only assess if it complies with emission legislation but also if the air quality of the places around the proposed activity can accommodate the impacts (immissions control).

For several other sectors, including noise and water quality, there is similar legislation with direct consequences on the location of activities. This sectoral legislation contributes to the physical separation of incompatible activities in the land use plans.

Environmental aspects are also considered by the municipal land use plan through two other sources of regulation: the national agricultural reserve (RAN) and the national ecological reserve (REN). There are strong limitations on the uses of land that can take place on either of these reserves. Every municipal land use plan has to delimit the prime agricultural land, that by law is preserved from urban uses and other aggressive activities. The REN includes the most environmentally sensitive areas of a municipality, including very steep slopes, areas of high infiltration, biotopes, etc.

In Portugal, during the nineties, there has been a significant increase in the amount of legislation to protect the environment and to integrate environmental concerns in land use planning. Additionally, many municipal land use plans and other local physical plans have been developed. These plans may prevent future conflicts between incompatible land uses, sometimes at a price of too much spatial separation and segregation, creating new needs for transportation and mobility, which in turn will have some environmental impacts.

However, additional questions remain. How is this new wave of land use plans performing in relation to already existing land use conflicts? What special planning methods have been applied, under which conditions and with what success? Who pays for the implementation of the solutions?

One of the planning tools to address this problem, frequently inherited from previous times, is land use management and the restructuring of existing uses, taking into consideration their level of compatibility. Three case studies will be presented with the aim of providing some clues. These cases reflect the current practice of municipal planning approaches to issues of conflicting uses within the Metropolitan area of Lisbon.

5.3   Case Study A: The Cement Plant of Alhandra

The privileged location of Alhandra, on the banks of the navigable Tagus river and very close to the main railway line as well as to the first highway that serves Lisbon, contributed to the early attraction of enterprises to this place. The cement plant was developed in the beginning of the 20th century and progressively became one of the largest industries of the area.

The village of Alhandra was quite small at the beginning of the 20th century, with a tradition of fishing and agriculture. The village and the cement plant were not adjacently located. The urban area grew in all directions, due to its closeness to Lisbon and partially influenced by the factory. On the other hand, the plant expanded in the direction of Alhandra. Nowadays the village and the plant are next to each other.

The cement plant has two furnaces which respectively produce about 160 and 110 tons per hour of clinker. The plant has a yearly capacity of two million tons of cement and consumes around 1,000 tons of coal per day. It currently employs 300 workers and in the nineties it became owned by the state. The village of Alhandra has a few thousand inhabitants.

Key Conflicts

There are several important conflicts resulting from such a huge factory operating close to an environmentally sensitive land use. The main negative effects are air pollution, noise, visual intrusion, water pollution and spillovers from heavy traffic generated by the cement transport.

The air pollution is immediately perceived by any visitor. One can breathe the small dust particles in the air. Its effects are quite visible on the roofs of the buildings, turning the typical red colour into a grey one. The cars parked on the streets quite often are sheltered by a special car cover in order to protect them from the dust.

In spite of the heavy investments in pollution control devices, over 10 million euros in the years before the Expo '98, the amount of particles in the air still sometimes exceeds the level of 120 mg/m^3 recommended by the World Health Organisation. The air quality has improved, but still remains at unacceptable levels.

The population of Alhandra perceive the air pollution as a deep problem affecting the quality of their lives and there is opposition to the continued operation of the factory. On the other hand, the inhabitants have lived under these conditions for decades and many families get their income from the plant.

Noise is generated by the factory during the cement production process. It is perceived mainly at night and in the residential areas closer to the plant. The cumulative effect of the noise from the factory and from the railway and highway nearby the residential areas are contributing to make the situation worse.

The visual intrusion of the factory is significant, since the plant with its large grey building and high chimneys can be seen easily from every point of the village. The large parking place for the cement trucks also produces a very negative visual effect. One side of the parking with dozens of huge trucks faces the factory and the other side faces small residential houses.

The cumulative effect of all these different polluting factors creates a very poor environmental quality with significant negative impact upon the people's quality of life. These effects are reflected in the urban decay of the surrounding areas.

Municipal Planning Approach

At PDM Level: The studies that support the municipal land use plan identify the cement plant as an important source of air pollution resulting in complaints by surrounding residents. However, the zoning contained in the land use plan does not change the existing land uses but only confirms these uses. The area of the cement factory is zoned as an industrial use for classes A and B, which permit heavy manufacturing. The residential area is zoned as urban area.

The responsible people for the PDM justify this situation, saying that the scale (1/25,000) of the PDM is too general to deal with this problem. According to the same source of information, the adequate planning level should be a detailed plan, at the 1/2,000 scale.

At the Detailed Plan Level

A detailed land use plan does not yet exist for Alhandra. The municipality is planning to formulate one in the future, which will probably consider the pollution problems. In the planning department there has been some discussion of creating a green area between the cement plant and the residential area, connecting the Tagus waterfront with the open space and giving access to the small mountains nearby. Presently, there is a shortage of open space between the two incompatible land uses. However, it could gain some area by pulling down some unused and rundown warehouses.

According to the planning team, there is a lack of technical and planning methods for identifying the essential width of such a green space, which would take into account the cumulative effect of the different kinds of pollution.

At the Building Permit Issuing Stage

The building permits for the residential area have been given only on the basis of architectural and urban planning criteria, i.e. taking into account the height and dimension of the buildings nearby. Environmental aspects are not considered in the decision.

This has resulted in a few buildings being erected in areas subject to very high air pollution. These buildings are easily detected by the red colour of their roofs. A few more projects are being built, including a building for social activities of the church and a building for residential purposes.

At the Air Quality Control Level

The municipality has created a laboratory for air quality control, focusing on dust particles. When the pollution levels exceed the recommendations made by the World Health Organisation, the samples and the report are sent to the cement plant. The reports are also sent to non-governmental environmental organisations operating in the area and to the Ministries of Environment and Industry. So far, there has been no practical intervention following this up.

The municipality has a very limited legal control over the operation of the factory, as permitted by the Central Government, so pressures to reduce emissions are quite limited. Negotiations have been the chosen way to deal with the problem at the municipal level. There is some compensation given by the cement plan, including support for the laboratory for air quality control, and these initiatives seem useful but still insufficient.

Conclusions

Due to its scale, the existing municipal land use plan (PDM) does not seem to adequately deal with existing land use conflicts. Detailed plans are better fitted for this purpose, and not yet developed.

After large investments to control air pollution at the source, an unacceptable level of pollution still remains in the residential area. Costs and technical solutions will limit further significant improvement in the near future. Thus, there is a clear need to act also at the land use development and control level.

At the moment, there is available information concerning the levels of air pollution in the residential area. However, the additional step to interpret this information and act at the land use level has not been made. Local officials consider the pollution as exclusively of the responsibility of the factory, and intervene only in controlling emissions instead of also restructuring the overall surrounding urban area.

There are important questions related with the social impacts of the environmental zoning. Should a part of the residential area be stopped to growth or even be pulled down? The existing social fabric of the village could be much damaged in this attempt to change Alhandra into a cleaner and healthier place to live.

5.4   Case Study B: Alcântara Industry

Alcântara is a district of Lisbon located in the west side close to the river. This urban area had a diversified industrial structure at the beginning of the 1980s. During this decade, the City experienced a considerable change, with a fast reduction in industrial employment, from 95 thousand jobs in 1981 to about 70 thousand in 1990 (PDM 1993, p. 86).

This resulted from the bankruptcy of firms due to changes in competitiveness, as well as an increase in the land and building costs which pushed bulk industries to the periphery. Though some of the industries located in Alcântara did not survive this trend, two of them are still active: SIDUL – processing sugar and Quimigal –

producing chemicals (Marques 1992).

Alcântara presents nowadays a mixed use urban space, within which housing, services, commerce and still these two important operating industries, live side by side. The industries, a legacy of the industrial development of the 19th century, were built to supply jobs and soon were responsible for the appearance of housing for workers in surrounding areas, which continued to grow as the town expanded.

Moreover, Alcântara is also an interface of several modes of transportation, including two passenger train lines interchange and one of the port railway connection. This generates heavy vehicle and pedestrian traffic.

A preliminary survey revealed that about 97 per cent of people crossing the area identified pollution, particularly noise, as the main nuisance. Though vehicle traffic strongly affects the local environment, the pollution generated by the industrial activity contributes a considerable share (Marques 1992). Local air quality becomes even more serious here since the Alcântara valley, due to its configuration, prevents frequently air circulation. The two industries (SIDUL and Quimigal) currently in activity, throw gas effluents to the atmosphere, easily identified by its strong unpleasant smell. This nuisance is particularly noticeable in the area surrounding the industries plant facilities and in the evenings (Marques 1992).

Industries also have a considerable visual impact, mostly due to their bulk and inattention to physical infrastructure maintenance, representing a serious impact to anyone crossing the area. This is even more serious because of a virtual absence of open green spaces which could provide for some amelioration.

Key Conflicts

One of the most important conflict of uses in Alcântara is the closeness of this polluting industrial activity operating close to the residential area. Essentially an industrial area, when it was created, this space soon became residential due to the need to supply housing for the workers which generated a fast growth of housing in its vicinity.

This was a result of the industry operating as a pole of attraction, but was mostly due to an almost total absence of planning policy for the area. Although growth of the industrial activity has slowed down lately, the two industries under study are still operating and present a situation which is calling for an urgent intervention.

Municipal Planning Approach

This analysis will focus on the way the Lisbon municipal planning process addressed the specific issue of conflict of these two sizeable industries inserted within a residential area. Three levels of planning within the local level will be considered: the Strategic Planning[4] a peculiar approach to planning followed only by a few municipalities, the Municipal Plan[5] and the Priority Projects and Plans[6] legally supported by Detailed Plans.

At the Strategic Planning Level

The Strategic Plan is general in character and provides a statement of intentions. Several proposals in this strategic plan provide some context of intervention, namely the rehabilitation of housing, the supply of leisure spaces including green areas and public spaces, improvement of the town environmental quality, rehabilitation of physically deteriorating structures and considering the possibility of alternative uses, improvement of accessibility, and amelioration of congestion through improvement of the interfaces between modes of transportation. These proposals show concern for restructuring parts of the town, in addressing physical deterioration and in providing solutions for congestion.

Thus the setting to address these issues has been created, and more specific strategies at the PDM level are proposed.

At the PDM Level

The PDM addresses a more detailed level, and Alcântara is mentioned within the urban planning objectives (PE/PDM 1990, p. 34) aiming for the modernisation of the Lisbon industrial structure through reconverting declining industries, creating technological areas and areas of integrated activities, but most important the preservation and assurance of balance of housing with other local uses in the existing neighbourhoods.

These industries, Sidul and Quimigal, belong to the waterfront, designated as strategic areas for urban intervention (PE/PDM 1990, p. 35). The objectives include: the preservation of the cultural aspects and their integration in improved urban spaces, the betterment of the living conditions of the residents through the rehabilitation of housing, and the reorganisation and renewal of industrial areas, with the objective to introduce new functions (PE/PDM 1990, p. 35). In the proposed Rules Report associated with the PDM, these two industries are located in the Area of Mixed Uses for Urban Reconverting, which in accordance with the Land Uses Classification Map, specifies the location of mostly residential and tertiary uses (PDM-Rules 1993, p. 59).

This implies among other requirements the need to relocate existing uses incompatible either with new uses or with the proposed urban solution (PDM-Rules 1993, p. 62). In fact, the Operative Units of Planning and Management[7] proposes, in the Specific Land Use Plans for Alcântara Rio, mixed uses (residential, tertiary and compatible industry), with a minimum of 40 per cent of the area to be used for housing (PDM-Rules 1993, p. 105). Moreover, future development has to be consistent with the uses in surrounding areas, to assure the urban integration of the transportation network and to preserve buildings and open spaces[8] (PDM-Rules 1993, pp. 105-6).

At the Detailed Plan Level

Detailed Plans provide specific proposals at the local planning level, and suggest the implementation phase by setting guidelines for the specific urban development for the area. The planning area includes the space along the Alcântara river.[9] The stated strategic objective is to 'make the City turn to the river' (PE/PDM 1990, p. 54). This implies redeveloping the industrial zone, with the rehabilitation of buildings with high quality, the transformation of local uses from industrial to tertiary and even quaternary, and increasing housing. All this has to take into consideration integration of this area with the surrounding urban areas.

The type of intervention proposed is quite ambitious. This is revealed by the negotiation and agreements already underway with the local owners to carry out the urban recovery, the expressed intention to seek international proposals for the development of the Local Urban Plan, and to carry on joint Specific Plans and Projects with local owners' associations to be developed by private enterprises (PE/PDM 1990, p. 54).

The Detailed Plan for the area where the two industries are located will be developed within the following boundaries: Rotunda de Alcântara, 24 de Julho, Infante Santo and Industrial Fair of Lisbon.[10] This area will be planned mostly for housing and tertiary uses.

The Municipality has been negotiating with local owners of large parcels to participate in converting the uses in the area either through relocation or adaptation of activities. So far, the Sidul has already shown interest in transforming the present type of activity, and the Quimigal is willing to relocate. It is still early to draw conclusions whether these changes will take place. However, restructuring of the area to residential and service use is under way.

Conclusions

Several conclusions can be drawn from this case:

1   there is a *commitment* at the different levels of the Lisbon local planning process to address the problem of conflict of use. This will work only if these guidelines are expressly included in the detailed plan for the area;
2   therefore, the several levels of the planning process offer effective means for intervening, by setting the framework for intervention and also by specifying the type of action to be followed locally, providing *institutional coverage*;
3   there is commitment to effective implementation of the planning guidelines. This is particularly apparent in the *negotiations* with local owners, to resolve difficulties that can arise and to get them involved in the process;
4   throughout the whole planning process, there has been strong *political support* from the Presidency of the municipality, which was responsible for setting in motion the local planning process;
5   moreover, a *coherent structure,* from the more general to the specific levels of intervention, has been developed and followed;
6   there is a strong commitment to restructure the area for mixed *uses*, predominantly residential and tertiary, but with some allowances for compatible industries;
7   besides this objective several *overall targets* for the area have been set, including the urban integration of the transportation network, re-establishing

the linkage to the river, the preservation of buildings of quality and open spaces, and integration of this planning area with the surrounding areas;

8   special negotiations with the owners of the two industries (Sidul and Quimigal) have been carried out and new developments will be under way (e.g. relocation, use transformation).

In summary, it appears that the more general planning levels have provided an adequate context for intervention. It is still early to analyse outcomes, since the specific solution can only be assessed once the Detailed Land Use Plan for the area is developed and approved.

It is however reassuring that objectives and planning guidelines are proposed to restructure the present urban space, taking into consideration issues directly connected with environmental quality and applying planning devices to address them. But even more important is the special concern for effective implementation through the efforts being developed to involve agents and financial resources and to closely follow the process. This commitment and concern by planning at the municipal level for redevelopment as a residential area and for local environmental quality is unusual.

5.5   Case Study C: Expo '98

The Intervention Zone (I.Z.) in the eastern portion of Lisbon for the Expo '98 occupies an area of about 310 ha. The location was chosen in 1992[11] by the International Exhibition Office. Portugal was selected as the host for the last International Exhibition of the 20th century (June 1998) being given priority under the Toronto proposal.

The Expositions before Expo '98

An exhibition such as this has an important impact on the urban environment. Similar expositions have created landmarks with profound repercussions for the History of Urbanism and Architecture of the countries which have undergone this experience. These impacts include for example: the introduction of new materials used in construction, new concepts in the approach to the problems concerning society at that time, and the contribution made to the development of the utilised technologies.

At the end of the 19th century the 'architecture' and the 'materials' used were innovated through holding the exhibition, such as the application of metal and glass brilliantly handled in the Crystal Palace project in London in 1851, the use of iron in the Eiffel Tower in Paris in 1898, new ideas demonstrated in the Espirit Nouveau Pavilion by Corbusier and in the USSR by Melnikov, or even in the recently rebuilt pavilion (1992) by Mies Van der Rohe (Barcelona 1929), which is considered a historical turning point in Modern Architecture.

By the middle of the 20th century, the universal and the international exhibitions were characterised by a broader impact, leaving visible changes in the design of the city. This allowed multiple initiatives to totally renew parts of the urban structure. It was not only the architecture or the material employed, but also the concepts and philosophy which underpinned the architecture and gave rise to the 'model' of development chosen for the city.

Important landmarks in the expansion of the city of Seville, such as the Plaza de España and the Parque Maria Luisa planned by Forrestier for the Hispanic-American Exhibition of 1929, can easily be seen as a result of this. Brussels experienced a major redevelopment with the 1958 Expo. After 1960, Osaka became an extension of Tokyo, a new satellite city with good accessibility and major investment in the travel and transport networks as well as in the overall infrastructure.

Recently and much closer to Portugal, the Seville Exhibition (1992) has been remarkable on every level for the profound alterations in redefining the city's urban space and re-equipping it with an improved level of infrastructure and facilities.[12] However in the case of Expo '92 in Seville, difficulties occurred related to the use of the exhibition facilities after the event. This could have been a consequence of the unexpected economic crisis and failure in developing scenarios for the future.

Having in mind these limitations, planning in Lisbon has sought to regenerate the city and to make development in it profitable, having abandoned methods of indiscriminately creating big expansion areas where, in many cases, there was little interest by private firms to undertake development that would implement the plan. In effect, these proposed expansions were often restricted to designating land uses and were inattentive to requirements for financial success for the developer or for the larger economic situation.

The Environment and the Expo '98

Toward the end of the 20th century, the task of addressing environmental problems has become a central issue and a priority for our future. Additionally, the oceans and their resources represent a disputed set of issues among the nations and an essential element for the survival of mankind. In this context, the choice of one of the most rundown environmental areas of Lisbon,[13] as well as the chosen theme: 'The Oceans, a Heritage for the Future',[14] are a demanding challenge for all those taking part either directly or indirectly in this effort.

In order for planning for this event to be successful, one needs to have a thorough understanding of the Intervention Zone, a clear definition of the project, to search alternative solutions, to analyse the state of the environment in this zone, and to foresee the effects that this project may cause that environment, as well as to consider the eventual difficulties for implementation simultaneously. The Environmental Impact Study is an important instrument in this situation, and in addition to conventional content will need to include the compatibility among the various participants and agencies involved in order that the long and short term objectives of the project are met. The environmental safeguards will have to take into account both the economic and the environmental aspects, which will need to converge as decisions are taken, while avoiding errors difficult to correct, given the limitation of available resources.

The project will not succeed if decisions focus on the short term requirements of the event, and overlook future uses of the Expo site. To do so would result in wasteful investment in infrastructure and political resentment. Yet, the short time line for making decisions mean that thorough research into existing conditions and analysis of options will not be possible. Consequently, reliance will need to be placed on working towards consensus among the decision-makers involved, within the framework of the objectives for the project and a set of evaluation criteria.

The International Exposition 1998 (Expo '98)

The location within the city The Expo is located in a vast, rundown, urban riverside area that makes part of the 'forgotten' city: the Eastern Zone. Unlike this new site, the zone chosen for the 1940 Portuguese World Exposition is Belem-Jeronimos which contains monuments identifying it as the place where the great Portuguese voyages of discoveries set off from in the 15th and 16th century. This area of the earlier Expo was also preferred for the court residence in the 18th century, and by the Republican regime in the 20th century.

The lower city was rebuilt after the 1755 earthquake by the Marques de Pombal. Terreiro do Paco, part of this reconstructed area, is an innovative work of 18th century urbanism and represents the heart of the city, which has expanded north.[15]

For decades, the eastern zone has been a neglected area. About one sixth of the city's area, previously in farms and estates, was expropriated in the 1940s, and up until the nineties the state and public entities have built three of the biggest social housing projects, including Olivais and Chelas, with a population of around 120,000 inhabitants. These developments are located in the riverside area, near the Expo. This area is bounded by the railroad line and the Tagus river, comprises 310 ha, and is located in the municipalities of Lisbon and Loures.

Key Conflicts

This area is relatively isolated from the centre of town by natural barriers and the railway line. The building of a railway station in Santa Apolonia and the existence of local docks has made it the location for industrial as well as other economic activities, such as large units of oil refinery production, warehouses belonging to small and medium scale industry, and container storage. There are also an abandoned main wharf of about 10 ha built in 1940. Other activities located here include an old industrial slaughter plant, large areas of degraded housing and shanty towns, harbour facilities, storage of petroleum derived products, storage of scrap war material, concrete factories and unloading of sand-dredgers, a cold-storage depot, a Lisbon Council waste disposal treatment facility, a sewerage treatment station, a sanitary landfill, and a solid waste treatment station. In addition to the soil and air pollution caused by oil derived products from the refineries and petrochemical industry, as well as the landfills, we are still facing a serious problem of significant pollution in the Trancão river.

The river watershed is managed by eight municipalities, and waste waters from their combined population of 1.3 million people drain directly into the river. This waste is also produced by farming (pig and fowl breeding), food and drink industry (fish and meat canning), chemicals (pharmaceuticals products and paints), graphic printing and wood and metal industry. Dealing with this pollution is an essential priority since this river is an open sewer.

Main Actors

Within the geographical area of the Expo '98 site, there are three governmental entities with development and municipal responsibilities: the Lisbon City Council, Loures City Council, and Lisbon Harbour Authority. Additionally, the Regional Authorities and Central Administration also have power of intervention. For the Expo '98, the government enacted special legislation giving exceptional powers to a company specifically created for this purpose: Parque Expo S.A.[16] The zone, 'Critical Area of Recovery and Urban Reconversion', is defined in this legislation, and the Parque Expo S.A. has responsibility relating to its conversion.[17] Thus many of the powers are taken from the local authorities who usually oversee and authorise local initiatives.

In the Municipal Development Plan (PDM), adopted earlier, the area is defined as an Operative Unit of Planning and Management (UOP 28) which will be covered by a specific Overall Plan, to be developed with Lisbon Harbour Authority, Parque Expo, C M Lo and other agencies having responsibility in this area. In February 1994, the Lisbon City Council approved the elaboration of an Urbanisation Plan for the area around Expo '98 (proposal 36/94). Even though this framework for planning has been established, it was unclear how this set of efforts will actively work to produce an integrated process involving all these key participants.

The Physical Features of the Expo Project

In the Parque Expo direct intervention area, a number of projects have already proposed.

Inside the site These include a sporting pavilion, auditorium, aquarium, multi-purpose pavilion, the Portuguese pavilion and 80 exhibitors and other groups with free-standing buildings which in after the Expo will be used as head-offices, as well as a range of buildings for restaurants, retail and bars. There will also be 'an internal system of transport, highly functional, innovative and of a certain spectacular nature'.

Along the river front Facilities will include a quay, bridges, canals, wharves and inland ponds or lakes, as well as a boat station. This portion of the city is intended to become one of the most accessible in Lisbon, through the city's road network, two planned ring throughways, the second bridge over the Tagus river, new boat crossings, subway connections, an extension to the bus network, and direct links to the railway stations and to the airport. In addition, there will be a direct link to the North-South highways, and a 20 ha car-park which is to be converted into housing and office facilities after the Expo. In order to accommodate all these physical developments, a project is being drawn up to build an 'inter-modal transport platform', to be named 'Eastern Station'. Next to this transportation centre, local hotels, housing, retail and office facilities will be built.

The pollution of the Trancão river A huge effort which includes the river bed to the North, will be made to tackle this.

North urban park In the northern zone a park is to be built with sporting facilities. In order to accomplish this, the rubbish dump must be recovered and integrated into the water treatment plant.

Conclusions

Redevelopment of the Eastern Zone and development of the site for Expo '98 is a large and complex planning project. Plans include relocation of the current sources of pollution, in all 94 public and privately owned facilities. These include oil tanks, slaughter houses, refineries and treatment plants, which need to maintain their levels of service since these are essential to the functioning of the city.

      These relocations were not studied prior to June 1992, when the decision to hold Expo '98 here was made. Also, several of these facilities have recently made large investments to improve their operations. As yet, no studies have been undertaken to assess the consequences of relocating existing uses which produce pollution, nor how to minimise their impacts on the areas to which they will be moving.

      It is certain that the Eastern Zone will be radically changed by preparations for Expo '98. A concentrated investment of more than 60 million euros will be made during a period of only four years. But major questions remain, including the returns on this investment for the city, the region, and the country.

5.6   Final Considerations and Suggestions for Future Research

Several lessons can be drawn from our review of these cases:

- the planning process should be *flexible*. The traditional rigid plans have to give place to more adaptive types of plans, which consider the context of the area under consideration, and for using several kinds of plans. Policies have to be translated into objectives, in the present case through the Strategic Plan (PEL). These objectives have to be converted into rules which are as quantifiable as possible by the Municipal Plan (PDM), and then it is essential to give territorial expression to those rules with the Plans and Priority Projects (PPP);
- throughout the process, but primarily at this last stage, the *involvement of the agents* who will be engaged in implementation, and the assessment of resources, play a key role. The municipality has a key function in facilitating the negotiation and implementation process, providing not only the necessary arenas for discussion but also the following up of the process;
- an important part of the decisions dealing with resolving land use conflicts is concerned with *social impacts*. Measures should not be adopted that adversely affect the daily life and neighbourhood connections of local residents. This requires effectively listening to the persons affected by the process;

- the *Detailed Plan* is a useful means of expressing spatially the policies and objectives from higher levels of planning, and is the main vehicle for carrying out task of providing operational solutions to conflicts of use. However, this task must be supported by a coherent body of policies and objectives at the other levels of planning to assure that these aims are implemented as intended;
- the Units of Management and Planning, into which the municipality of Lisbon is divided, provide a *way to implement the planning process* by defining the specific guidelines for each of the sub sections of the town, and thus establish the basis for developing Local Land Use Plans;
- this framework for planning provides *support for conflict resolution*, making possible to restructure incompatible uses either through relocation or change of use, through involvement of owners in reaching a negotiated solution;
- intervention should start by acting at the polluting source. When this is insufficient, then relocation or transformation of uses must be considered.

Until the beginning of the nineties, most of the planning process has been concerned with projecting and planning for newly occupied areas. At the stage that most space has been occupied in our metropolitan areas, and that the existing development is suffering drastic changes contributing to their decline, it is time to shift our focus to recover the living town to its residents through creative use of available planning tools. For this new concept, modelling the urban spatial structure must be tried to target comprehensive restructuring with simultaneous assurance of local identification and quality.

A key aspect of intervention is the way conflict of uses generated by urban expansion can be addressed. Though important, it is not enough to identify sound objectives. One must go a step further, translating these objectives territorially and identifying the mechanisms to assure their implementation.

It is obvious from the way that planning has been evolving that planners became aware of the inadequacy of rigid plans. There is a growing concern in the implementation process to assure that the intervention which is proposed in the plan is the product of wider participative mechanisms, including negotiation with the decision makers and the local investors. Moreover, there is a commitment to implement objectives by quantifying them as much as possible. These are two of the features missing in most of the traditional plans used in Portugal. These plans represented great intentions but were insufficient in providing tools for implement them, because they did not adequately consider the context of implementation, and/or they did not adequately specify the targets to be attained.

Notes

1	João Farinha is Assistant Professor at the Department of Sciences and Environmental Engineering, New University of Lisbon, Portugal.
2	Lia Vasconcelos is Lecturer at the Department of Sciences and Environmental Engineering, New University of Lisbon, Portugal.
3	António Perestrelo is Lecturer at the Department of Sciences and Environmental Engineering, New University of Lisbon, Portugal.
4	PEL, *Plano Estratégico de Lisboa*.
5	PDM, *Plano Director Municipal*.
6	PPP, *Planos e Projectos Prioritários*.
7	Unidade Operativa de Planeamento e Gestão (UOP) - Lisbon Municipality according to the PDM is divided into 29 UOP, this specific area belongs to the UOP 19.
8	Calvário, Alcântara and Junqueira.
9	Space located inside the following lines: Avenida de Ceuta, Rotunda de Alcântara, Avenida 24 de Julho e Avenida da Índia.
10	Industrial Fair of Lisbon (Feira International de Lisboa).
11	The same date as the Rio de Janeiro Conference which produced 'Agenda 21' – a document which analyses and gives special relevance to the theme of this Exhibition 'The Oceans'.
12	'The concepts and processes of systematic intervention in the existing city have evolved significantly throughout recent decades and, in consequence, have changed the intention and tools of Planning as well as the organisation of management'. (Nuno Portas – see Sociedade e Território, no. 2).
13	With obsolete industry, scrap metal dumps, air, soil and water pollution with waste residuals, in addition to low-income housing and degraded buildings in peripheral area.
14	'A quarter of a century after the International Exhibition of Okinawa (1975) that shared this theme in an optimistic manner, anticipating the future: 'The sea we want to see', this did not predict the intensive exploitation of resources. Expo '98 aims to reflect this and bring to the world's awareness means to avoid the accelerated destruction of a formidable source of life and resources that exist in these vast waters that cover the planet and that are crucial to the global ecological balance...'.
15	700, 000 inhabitants (1991).
16	For reference see DL 87 and DL88/93 which define the intervention zone and the legal basis of the company whose capital is completely public.
17	D/L 16/93.

References

Câmara Municipal de Lisboa (CML) (1992) Direcção de Projecto de Planeamento Estratégico, Plano Estratégico de Lisboa, CML, Outubro 1992.

Câmara Municipal de Lisboa (CML) (1990) Plano Estratégico/Plano Director Municipal, Proposta de Objectivos, Bases, Metodologia e Calendários, CML, Agosto 1990.

Câmara Municipal de Lisboa (CML) (1993) Plano Director Municipal, Regulamento (não aprovado), CML.

Câmara Municipal de Vila Franca de Xira (CMVFX) (1993) Plano Director Municipal – Regulamento CMVFX.

Garrett, C. (1993) A Integração de Critérios de Qualidade do Ambiente na Elaboração de Planos Directores Municipais; UTL, Lisbon.

Jornal Expresso (10 Jul 93).

Marques, Dâmaso Silva, Margarida Rosado, Sofia Mourato, Teresa.

M.Madeira (1992) Planeamento Urbano, Conflitos Urbanos em Alcântara Março 1992.

Parque Expo '98 'The Oceans – a Heritage for the future'.

Parque Expo '98 'The Challenge and the Opportunities'.

Parque Expo '98 Boletins da Expo (1-8).

Parque Expo '98 'Project Management'.

Pimenta, Carlos (1993) 'Conferência sobre Ambiente', FCT-UNL Junho 1993.

Pinho, P. (1991) A Preservação e o Controle da Qualidade do Ambiente na Elaboração dos Planos Directores Municipais; CCRN, Porto, Portugal.

Revista Sábado' (1992/93).

Revista da Ordem dos Engenheiros (Jan/Dec 93).

Revista da Associação dos Arqui tectos Portugueses – AAP (130).

Revista Sociedade e Território No. 2.

Acknowledgement

Special thanks are due to all the people contacted who kindly have provided the information without which this work was not possible.

Chapter 6

A Noise Remedy Program:
Seattle-Tacoma International Airport

E. Munday[1]

6.1 Historical Introduction

In the late 1960s and early 1970s many people across the country began suing airports for damages caused by aircraft noise. This also happened in the Seattle area. Because of this, in the early 1970s the Port of Seattle (owner of Sea-Tac Airport) and King County (the county in which Sea-Tac airport is located) performed a land use study called the Sea-Tac Communities Plan. In this plan, 1,008 homes were identified to be purchased by the airport and removed and 770 homes were identified to be provided 'Purchase Assurance'.

The Acquisition Program began in 1974. In the late 1970s, the Federal Government enacted the Part 150 Program which provided that airports which had a noise abatement/mitigation plan approved by the Federal Aviation Administration (FAA) would be eligible for federal funding at the rate of 80 per cent of the cost of implementing said plans.

In 1985, Sea-Tac had such a program approved by the FAA. We were about the fifth airport nation-wide to have a Part 150 Program approved. Sea-Tac's Part 150 added 361 homes to the Acquisition program and identified 9,867 (including the 770 identified in the early 1970s) homes to be insulated, and in some cases sales assisted.

6.2 Acquisition

The areas identified to be acquired in the early 1970s were based on 'extended clear zones'. This was done because no standards existed at that time to identify acquisition based on noise levels. These clear zones were based on the likelihood of aircraft crashes when landing at or taking off from airports.

When the acquisition area was expanded in 1985, the expansion was based on *noise levels* using the FAA supplied Integrated Noise Model (INM) computer model of average annual noise levels. The levels identified by the FAA as being eligible to be acquired are those above 75 DNL (Day Night Level).

The FAA has partially funded all of these acquisitions. Whenever federal funds are used for an acquisition program, it is required that the acquisitions be performed according to the 'Uniform Relocation Assistance and Real Property Acquisition

Regulations for Federal and Federally Assisted Programs' (49CFR Part 24). These regulations specifically detail how the acquisitions are to be performed and what relocation benefits must be subsequently made available to the displaced residents.

Acquisition Procedures

The homeowner is notified by certified mail that his house will be acquired. A Port staff member then meets with the homeowner at his house to explain the process. An appraisal is then performed by a Port hired appraiser. The house is appraised as is/where is except that minor maintenance problems are not deducted from the appraisal value. The appraisal is based on *actual* sales of similar houses located in similar neighbourhoods. Adjustments to the actual sales are made based on differences between the house sold and the house to be acquired. The appraiser determines those adjustments based on experience and appraisal industry standards. The appraisal is then 'reviewed' by a Review Appraisal firm hired by the Port. This firm verifies that the appraisal was done correctly. The Review Appraiser then sets the Fair Market Value (FMV) for the property. The FMV is almost always the appraised value as determined by the original appraiser, but sometimes will vary, especially if there is a rapidly changing market.

When the appraisal report and FMV comes to the Port, a staff member, with the assistance of a local Real Estate Agent, finds comparable houses that are *currently* for sale (within a 15 mile radius whenever possible). Out of those 'comparables', the staff member selects the one that he feels is most comparable, i.e. similar size, similar number of rooms, similar construction, etc. The only exception to this is that the comparable house *must be 'Decent Safe & Sanitary'* (DS&S). DS&S is a requirement of the federal government which means that the house must provide a weather tight residence containing the accoutrements that would normally be found in a residence, i.e. heat, water, bathrooms, kitchen, etc. This means, in some cases, that the comparable house is a much better house than the house which is being acquired. The difference between the price that the comparable would be expected to sell for (i.e. what the seller might be expected to sell for, keeping in mind that most offers are less than the advertised price) and what we offer the homeowner to acquire their property is called a Replacement Housing Payment (RHP).

After staff determine the RHP, an offer is made to the homeowner in the amount of the FMV. The actual appraisal is never shown to the homeowner as it will be used in court if the property has to go through condemnation. At that time, the homeowner is also informed of how much the RHP will be so that he will know the total amount that would be available for him to purchase another residence. The homeowner is given 15 days to make up his mind whether to accept or reject the offer. If the homeowner accepts the offer, papers are signed and the property is acquired. If the homeowner rejects the offer, he must get his own appraisal, toward which the Port will pay $200 (Washington State Law). When the homeowner appraisal is turned in to the Port, it is submitted to the same Review Appraiser as the original Port appraisal was. The Review Appraiser then determines a new FMV which may be the same as the original FMV or may be higher. This FMV is then offered to the homeowner. If the homeowner rejects this second offer, the acquisition is turned over to the Port's legal department to proceed with acquisition by eminent domain (condemnation). Of the 1,400 properties acquired by the Port, only about 20 were turned over to the legal department, and all of them were settled short of an actual trial.

Once the homeowner accepts the offer, the sale of the property is closed through a normal escrow process. All escrow and selling costs are paid for by the Port. When the property closes, and the homeowner gets the money (FMV less any underlying liens), the Port takes over ownership of the property. The occupant of the residence, who may or may not be the owner, then has 90 days of free rent. If the occupant is unable to vacate within the 90 days, i.e. they are building a house, have children in school, etc., then they may be allowed to rent their house back from the Port for a fair market rental amount until they are able to move. The occupant of the house is eligible for several Relocation benefits. These include moving costs, closing costs for the replacement dwelling, utility connections for the replacement dwelling, any difference in interest rates for new loans (up to the maximum of the old loan balances), etc. The occupant is also eligible to receive an amount up to the RHP amount as identified earlier by the Port, but only if he actually spends that amount to purchase the new dwelling (in addition to the FMV) and the dwelling purchased meets the DS&S requirements. The amount of Relocation benefits varies, but has a current federal maximum of $22,500, except in unusual circumstances such as when the acquired dwelling does not meet DS&S requirements or when reasonably priced comparable dwellings are not available.

There are special benefits available when a business is being acquired and when the acquired residence is occupied by renters. Only the occupant is eligible for Relocation benefits, including RHP payments.

6.3   Insulation

The areas identified in which to provide homes with insulation against noise were based on the FAA INM computer model. Areas eligible for insulation are those between 65 and 75 DNL. Due to the anticipation that the noise levels at Sea-Tac would decline between 1985 and 2000, and based on the fact that it would take at least until the year 2000 to accomplish all of the insulation, the Port used predicted year 2000 contours to establish the boundaries for the acquisition and insulation programs. It was predicted that the contours would shrink by about 5 dB over that 15-year period due to the phase out of the noisier Stage II aircraft, even though total operations were expected to increase significantly. Therefore the 1980-contours for the acquisition and insulation boundaries were actually closer to 80 DNL and above for acquisition, and 70-80 DNL for insulation. Another reason for using the 2000-contours was so that we would not commit to insulating homes based on 1985 contours, and then when we got to them in the 1990s have them no longer eligible for insulation (the FAA will not fund mitigation for properties outside a current 65 DNL contour). The FAA has partially funded all of the insulation at the rate of 80 per cent up to this time.

The Port hired a consultant who designed insulation for 21 homes. The Port then hired one contractor to insulate those 21 homes. The Consultant also provided the Port with a computer program which could be used to design insulation for additional homes to be included in the program, based on relatively detailed data input by the operator of the program. The demonstration project was completed in September of 1987. The cost was about $500,000 for the consultant and about $500,000 for the contractor. The demonstration program was very successful except for insulating a mobile home. The noise reduction for the mobile home was insignificant. Based on that result, and later input by the FAA, we no longer attempt to insulate mobile homes.

The Port began the regular insulation program with the first contract awarded in December of 1987. The goal for the rate for insulating residences was 15 houses per month. That rate was achieved in 1990. At that time, based on the recommendations of the Mediation Agreement, the goal was raised to 30 homes per month. That rate was achieved in 1993. In late 1992, based on the Port Commission directive related to the planned development of an additional runway, the goal was raised to 100 homes per month. Effective January of 1994, the rate of insulation of homes at Sea-Tac became 100 homes per month.

Insulation Procedures

The program is publicised via newspaper articles, public meetings, airport newsletters, etc. Homeowners who think that they qualify for the program call the office or drop by. They are given an application form which asks for their name, address, when they purchased their home, and other miscellaneous information. The homeowner is also asked to supply a copy of a document which verifies when they purchased their home and, if available, a couple of pictures of the home.

Port staff then determine whether the house actually qualifies for the program based on a large-scale boundary map located in our office. If the house does not qualify, the homeowner is sent a letter of regret along with documentation which informs him of how he could provide his own insulation, should he so desire. If the house qualifies, another large-scale map which has contours overlaid on it is referenced to determine what the next highest nearest contour is to the location of the house. The data from the application, along with the contour, is entered into a computer database.

A computer program is then run which prioritises the houses. The current priority program considers three items. The first item considered is when the house was purchased by the current owner. One point is given for each year of home ownership since the year 1955. The next item considered is noise level (contour). One and one-half points are given for each DNL above 65. The third item considered is time on the waiting list. Two points are given for each month that the house remains on the waiting list. All points are carried out to four places. A printout of the list is made each month which shows all of the homeowners on the waiting list in priority order. The waiting list contained about 3,100 names in 1995.

A Pre-Qualification Packet is sent to each new applicant shortly after we receive the application. This packet contains information that the homeowner will need to prepare for the insulation process. Included in the packet is information about making sure that the title to the house is clear, i.e. that the person applying is legally the owner of the house. We have many applicants who have had problems with their title for varying reasons such as a death of one of the owners. Another item included is a Power of Attorney so that the homeowner can designate someone else to sign the insulation documents in the situation where the homeowner will not be available during the process, or if there are multiple owners who wish one person to sign for all of them. Another item is a subordination agreement. In order for the Port to provide insulation at no cost to the homeowner, the homeowner must provide an Avigation Easement to the Port. This Easement states that the property owner will not sue the airport for damages caused by aircraft noise or noise associated conditions. The Easement is filed with the county and is placed on the property title in perpetuity. If there is a large outstanding lien on the property, the lien holder must subordinate his lien to the Easement. This is so that if the property is repossessed by the lien holder, the lien holder cannot void the easement. Another topic is the Pre-Existing Avigation Easement. Some property owners around Sea-Tac won lawsuits in the late 1970s, and as part of the settlement of those suits the Port gave them money and they gave the Port an Avigation Easement. They cannot give the same easement twice, so they now have no Easement to give the Port in exchange for the free insulation. These homeowner therefore now have to pay the Port's share (20 per cent) of the insulation. Another concern is asbestos. The Port will not disturb asbestos during our insulation due to the liabilities associated with that disturbance. If the homeowner suspects that he has asbestos, he is instructed to contact the Port to have the material tested so that we can design accordingly. The last items in the Pre-Qualification Packet discuss general construction. If the house has major maintenance problems which would effect the viability of our insulation, they must correct those deficiencies prior to our construction. Also, if the homeowner wishes to do any remodelling which will effect the exterior shell of the house, it must be done either before we do our design, or after our construction is complete.

When the priority list is run each month, the top 130 homeowners on the list are sent a Homeowner Handbook and a letter informing them that they have reached the top of the waiting list and they should call our office to arrange to attend a Homeowner Briefing Session. The handbook contains background information about our program, an explanation of what acoustical insulation is, a checklist and flowchart of the insulation process, copies and explanations of all documents that the homeowner will see or be asked to sign during the process, a wrap-up summary of the program, and a list of names, titles, and phone numbers of all Noise Remedy staff members.

About 50 people representing 25 homes (usually both husband and wife) attend each meeting. At the meeting they view a 20-minute video about the program, receive a 30-minute briefing by a Port staff member, and ask any questions that they might have. At the end of the meeting, the homeowner is given an opportunity to sign an Initial Authorization which gives the Port permission to enter their houses, perform an inventory, and design an insulation package for their home. At the meeting they can also sign up for a specific time to have the inventory performed.

A Technician from the Port then goes to the house. The Technician walks through the house and identifies the construction materials, i.e. wood/brick/etc., number of doors, windows, rooms, etc. of the house. The Technician then inputs this data into a laptop computer. The computer contains a program designed by Wyle Laboratories of California. The program specifies the insulation to be installed in each room of the house. The program was designed based on information gathered through installing insulation on 1,200 houses in the Sea-Tac area in addition to information gathered by Wyle Labs over the years on other insulation programs in which they have participated. The criteria that affect the design are location of the house relative to the airport (i.e. how close to the airport and where the flight path is in relation to the house), how the house is constructed (i.e. masonry, wood, flat roof, two story, etc.), what type of heating system the house has (i.e. forced air or individual room heat), and what amenities the house has (e.g. fireplaces). The design as specified by the computer program can be modified by the technician based on his expertise and/or unique features of the house. We try to keep modifications to a minimum however. Jet aircraft noise reduction requires heavy materials due to the need to stop low frequency noise. The materials used in our programs include solid core exterior doors plus laminated glass storm doors, replacement Sound Transmission Class (STC) 44 windows or storm windows over the existing primary windows, R38 attic insulation, chimney dampers, additional sheet rock or sheet rock and sound board added to walls and/or ceilings, and a ventilation system so the homeowner does not have to open the windows to get fresh air. STC 44 windows are typically double or triple windows built with laminated quarter inch glass. The windows are not sealed units as the glazing must be spaced about two inches apart in order to achieve the sound reduction rating and no one makes a sealed unit with that much spacing. The windows are very expensive, about $60 per square foot (as compared to about $25 per square foot for normal thermal windows).

The ultimate design criterion is to achieve at least 5 dB of sound reduction in each room or achieve an interior noise level of 45 DNL, whichever requires the greatest sound reduction. We have found that the typical uninsulated wood frame house achieves about 25 dB of sound reduction and the typical thermally insulated wood frame house or uninsulated masonry house achieves about 30 dB of sound reduction before acoustical insulation is installed. Therefore we usually must achieve the minimal sound reduction increase (5 dB) in the Standard Insulation Area, currently 65 to 72 DNL, and up to 10-15 dB reduction in the Neighborhood Reinforcement Area, currently 70 to 80 DNL.

After the data is entered into the computer, a Scope of Work is printed from a portable printer. This Scope is presented and explained to the homeowner. A Standard Specification Book is also given to the homeowner. This book contains the specifications for all materials that may be used in our program, how the materials are to be installed, and hints as to how the quality of the installation can be inspected. After discussion of the scope, the homeowner is given the opportunity to sign the Scope or he can think about it and bring it in signed to our office at a later time. After the homeowner signs the scope, five copies are made and returned to the homeowner along with a list of participating contractors. In order for a contractor to become a participating contractor, they must attend a briefing on our program and then supply us with a number of documents including proof of insurance, a copy of their contractors' license, signed statements related to paying approved wages, non-discrimination, etc. There were 28 contractors on our list in 1995. The homeowner then chooses at least three contractors from the list to bid on his project. The homeowner invites the chosen contractors out to his house, provides them with one of the copies of the Scope of Work and allows the contractors to go through the house to determine their estimated cost. Any questions that the contractors or homeowner have are discussed between them at this time. Each contractor then leaves, writes up his bid to do the work, and submits it in a sealed envelope to the Port by the predetermined bid date that the Port has put on the Scope of Work.

On the predetermined bid date, the Port opens and reads the bids. The homeowner and contractor are welcome to attend that bid opening, however most do not attend. After the bids are opened, Port staff sends the homeowner a letter telling him who bid on his project, how much each bid was, and who the low bidder is. The Port will pay the amount of the low bid, however the homeowner may chose any of the other bidders to do the work. The homeowner is then responsible for paying the contractor chosen whatever extra amount the contractor requires to perform the work over the low bid amount. Usually the low bidder is chosen as the contractor who will do the work. Once the homeowner decides which contractor he wants to do the work, the homeowner makes an appointment to come into the office for a Final Sign meeting. At this meeting, the homeowner will sign the Avigation Easement, a contract between him and his chosen contractor, a letter of selection to the contractor telling the contractor to begin gathering the necessary materials and paperwork to perform the construction, and a Final Authorization which tells the Port that the homeowner agrees with the Scope of Work and wishes to proceed into construction.

The Port then sends the letter of selection to the chosen contractor. The contractor must then sign the contract, order the required materials, contact the homeowner to determine a Notice to Proceed (NTC) date (the date that construction will start), and supply the Port with proof of insurance for that particular project which includes the homeowner as a named additional insured. This insurance requirement is so that if a problem develops, the homeowner does not have to sue the contractor, but rather can go directly to the insurance company for compensation.

The contractor has 30 calendar days to perform the construction. During and at the completion of the construction, the work is inspected by the homeowner, or at the request of the homeowner, by Port technicians. The construction is done while the house is occupied. The contractor is required to leave a broom clean, lockable shell at the end of each day, i.e. each window or door must be removed and replaced within one day. After the final inspection, the homeowner is asked to sign the contract. The contractor then presents the signed contract to the Port along with various government required papers such as proof of wages paid, lien waivers, etc. The Port then makes a check out in the amount of the low bid and mails it to the contractor.

At the same time, the Port mails a questionnaire to the homeowner. This asks how the homeowner liked the process, the people and the results. It also asks various statistical questions. One page of the questionnaire asks specifically about the performance of the contractor. This page is photocopied and placed in a contractor's file which is available to all future homeowner to look at to help them decide who to select to bid on their homes.

6.4   Other Mitigation Activities

Sales Assistance

Sales Assistance is offered to all homeowners in the Neighborhood Reinforcement Area who have had their insulation completed. This assistance comprises establishing a Fair Market Value for the house based on comparable sales from outside the area eligible for insulation. In other words, houses which are not (or are less) impacted by airport noise are used to set the value of the house. The homeowner then markets the house through normal Real Estate channels for six months using any licensed agent and a county wide listing service. The agent must market the house in a manner specified by the Port related to price advertised and how advertised. If the house does not sell within six months, the Port takes over marketing and tries to sell the house for an additional two months. If the house still does not sell, the Port will buy the house for the amount that the homeowner would have received had the house sold for the full appraised value. If the house sells at any time during that eight month period, the Port pays the homeowner the difference between what he got out of escrow and what he would have gotten had the house sold for the full appraised value. Up until 1995 fifty-one houses were sold through this program. We have never had to purchase one of the houses. We have had to pay an average of 12.4 per cent of the appraised value in order to support those sales.

Purchase of Avigation Easements

If a property cannot be insulated, or if it does not make sense to insulate it, i.e. the property will be torn down soon), then the Port will consider paying the homeowner an amount to directly purchase an Avigation Easement. At this time, the only Easements purchased have been for mobile homes located on individual privately owned lots due to the fact that no one has discovered a method of insulating a mobile home satisfactorily. We have based the valuation of these easements on a document published by the FAA entitled 'Aviation Noise Effects'. One section of this document titled 'Effect of Aircraft Noise on Real Estate Values' examines a study done in the 1960s and comes to the conclusion that the average effect on Real Estate values is '[...] approximately 1 per cent decrease per decibel (DNL)' for each DNL over 65. The FAA has agreed to fund Avigation Easement purchases based on this percentage applied to the county tax assessed value of the property. That takes the Port out of the task of establishing an appraised value for each Easement that we wish to purchase. Up until 1995 we have purchased ten Easements at an average cost of about $5,000 each.

Insulation of Other Facilities

There exist about 60 Public Buildings within the Noise Remedy Program boundaries.

These buildings include public and private schools, churches, libraries, convalescent homes, fire stations, etc. There are also about 1,500 mobile homes that are renting spaces in mobile home parks and about 10,000 multi-family residences (apartments and condominiums). The Port is designing insulation programs to deal with all of these facilities except mobile homes. A program is being designed to assist mobile homes in being moved outside the Noise Remedy Boundaries. These programs began in 1994 and will cost about $100 Million to complete. Expected completion is shortly after the turn of the century.

6.5   Ending

The Port of Seattle, which operates Seattle-Tacoma International Airport, has been addressing noise impacts around the airport since the 1975. We have acquired 1,400 houses at a cost of $102 Million and are spending $115 Million to insulate 10,000 additional houses. We have plans to insulate other facilities at a cost of an additional $100 Million. The Insulation Program began in 1985 and has insulated 1,300 homes to date. The homes are designed and contracted on an individual basis. The current rate of insulation is 100 homes per month. All single family residential insulation is expected to be complete shortly after the year 2000.

Note

1   Earl Munday is Noise Remedy Manager at the Port of Seattle, Sea-Tac, Washington, USA.

Chapter 7

Land Use Conflicts at a Nuclear Weapons Site: Contamination, Industry and Habitat at Hanford, USA

J. Denkers and M. Power[1]

7.1   Overview

The 560 square mile Hanford nuclear reservation straddles the Columbia River in southeastern Washington State. For over 50 years, the area was closed to uses other than highly-secured nuclear activities. Several industrial complexes, employing thousands, brought large infrastructure investments on the site, and fuelled development of a small metropolitan area nearby. At the same time, large areas of the site were closed to any public access and maintained as buffer areas. These buffer areas have become significant preserves, in some cases the last in the Columbia River Basin, of habitat and other ecological values.

Cessation of nuclear weapons activities at Hanford, and a major focus on environmental cleanup and restoration, have led to intense discussions about future uses on the Hanford site. Proposed future industrial uses that will take advantage of existing infrastructure and workers collide with significant residual contamination and waste management problems. There are also conflicts among proponents of new industrial uses, agriculture, and habitat preservation.

This chapter discusses the development of the land use patterns at Hanford, and the emerging conflicts. Geographic information system (GIS) resources are used to illustrate and analyse the conflicts and issues. We also describe briefly the work of the Hanford Future Site Uses Working Group, an initial step, supported by the U.S. Department of Energy, the U.S. Environmental Protection Agency, and the State of Washington to provide policy guidance to help manage future site use controversies.

This chapter does not deal with one significant aspect of past and future land uses: American Indian uses and claims. We recognise the importance of treaty rights, rooted in past uses, and of legal requirements to protect cultural and archaeological sites. The subject is complex, and we are not able to display cultural and archaeological information at this time.

7.2   Past and Present Hanford Uses

The Hanford site was established in 1943. It contained the first large-scale facilities to produce plutonium for nuclear weapons, which began operation in late 1944. Industrial facilities were developed in three geographic areas. Nine nuclear reactors were built along the south shore of the Columbia River. Once irradiated, nuclear fuel was chemically dissolved in the '200' areas, in the centre of the site, to extract plutonium and uranium. Over time, five different separations plants operated in the 200 areas. Nuclear fuel fabrication, laboratories and support facilities were located at the southeastern corner of the site, near the City of Richland.

Production of nuclear weapons material ended in 1988. The principal activities at Hanford today are waste management and environmental cleanup. This includes decontamination, decommissioning, and removal of major facilities. The Washington Public Power Supply System operates a civilian nuclear power plant about eight miles north of the City of Richland.

There are approximately 1,400 contaminated sites resulting from nuclear weapons activities. Most are in the immediate vicinity of the industrial facilities mentioned above. However, approximately 200 square miles of the groundwater in the unconfined aquifer beneath the site are contaminated. The site has a road and rail network, power generating stations, and large-scale pumping facilities to bring water from the Columbia River for industrial purposes. Approximately 16,500 people are presently employed on the site or in directly related jobs in the City of Richland.

7.3   Use Conflicts

When the Army Corps of Engineers created the Hanford reservation, one of its great attractions was the absence of nearby population. Only a few thousand people lived within the site or nearby. Now, however, about 150,000 people live in the Tri-City (Richland, Kennewick, Pasco) area. The 16,500 Hanford jobs provide about one-quarter of the employment in the area, and 42 per cent of the payroll dollars. The second largest employment sector in the area is agriculture, accounting for about 12,900 jobs, but a payroll only one-fourth that of Hanford.

The community is highly motivated to find new activities that will continue the level of highly-paid, technical and professional employment prevailing in the area. In particular, community leaders see the availability of a trained technical work force, a large area of land well served by roads, rail, and power, and current national support for environmental cleanup as the bases for strong future economic growth.

Paradoxically, however, the fact that Hanford has been set aside from normal development and public use for the past 50 years means that much of it contains significant habitat values. There are many threatened or endangered species of fauna which nest or breed on the Hanford site, as well as threatened or rare plant species. Hanford contains the last significant parcels of Columbia Basin shrub steppe habitat. Congress will consider protection of the Hanford Reach of the Columbia River, essentially that portion within the site boundaries, for protection as a wild and scenic river.

Contamination from past operations poses potential conflict with both preservation and new development. For example, the 200 square miles of contaminated groundwater beneath the site limits both access to groundwater resources and activities that would develop a large hydrostatic head that would drive the contamination into the Columbia River. The intensity and the difficulty of waste management activities, and facility decontamination and decommissioning, make the '200' chemical processing areas unlikely candidates for new, non-waste-related uses, even though these areas have concentrations of infrastructure. Finally, the public perception of Hanford, when linked with that of other things nuclear, reduces the site's attractiveness for commercial agriculture, even though some areas farmed before 1943 remain uncontaminated.

Again, we note that Indian treaty rights, and cultural and archaeological protection, may add potential conflicts. These are not illustrated in this chapter.

7.4   Future Site Uses Working Group

In 1990-91, the U.S. Department of Energy (USDOE), the U.S. Environmental Protection Agency and the State of Washington agreed to establish a broad-based 'stakeholder' group to provide initial guidance and direction about future uses on the Hanford site. The principal need from the perspective of these agencies was to use the group's guidance to inform decisions about the timing and extent of environmental cleanup actions. However, the Future Site Uses Working Group, whose members worked through 1992, opened a broader regional discussion about future uses that continues in many forums.

The Future Site Uses Working Group included representatives of federal, state (including Oregon), local and tribal governments, local and regional economic interests, including agriculture and labour, and environmental groups, including Hanford-specific regional 'watchdog' organisations.

Briefly, three things are most significant about this group's work. First, they provided a broad-brush approach. While they divided Hanford into six separate geographical areas, they did not produce a land use plan in the usual sense of the term.

Second, the group did not make a recommendation of a single future land use. Instead, a variety of future use options were explored. However, a surprising degree of consensus was reached. The group came to recommend 'cleanup scenarios' for most of the site that maximise the options for alternative future uses. As a part of its process, the group visualised potential future uses for each of the six areas. All uses raised by participants that received the slightest amount of interest were carried through an analysis. There is no future use scenario for the reactor areas along the Columbia River that include either residential use or significant industrial use, even though the roads, railroads, power, and other infrastructure are present. Agricultural uses were proposed for only a small portion of the site.

Third, the Future Site Uses group recommended some principles to guide more specific future actions affecting land use. These principles include:

1   give high priority to protection of the Columbia River, particularly its water quality. Earlier cleanup planning, based on traditional risk assessment approaches, had not given the river such attention. The group's report signalled a growing recognition of the habitat resources represented by the Hanford Reach;

2   give high priority to groundwater. The report specifically recommended against use of surface water in such a way as to increase spread of contaminated groundwater;

3   concentrate waste management activities as tightly as possible to the central 200-area plateau. Recognise that this area would be used for waste management for the foreseeable future, but minimise the amount of land consumed for this purpose;

4   decisions about cleanup of the site and about future uses 'should result in decreased risk to public health and net benefits to the environment'. This 'do no more harm' was regarded as particularly important 'in light of the existing wildlife and plant life at the Hanford site';

5   clean up those areas that do not pose a particularly great hazard, but where removal or remediation of a relatively small amount of contamination would free the areas for other beneficial uses. This meant primarily the 'North Slope', presently managed as a wildlife refuge, and the Arid Lands Ecology Reserve.

In its work, the Future Site Uses group had some broad geographical information available to it. Since its report was completed, more specific data have been integrated into a geographic information system (GIS). This kind of information will be essential as responsible agencies and the public move to apply these broad principles to more specific decisions about cleanup and future land uses.

7.5   Issues and Conflicts Illustrated

Some of the issues and conflicts mentioned above can be illustrated by using mapped geographic based information. First, note that the two large areas with little development or contamination, the North Slope and the Arid Lands Ecology Reserve, are scheduled for early cleanup and for transfer from USDOE to other ownership. In both cases, habitat preservation is a high priority. Some area residents advocate using portions of both areas for agriculture. This is especially true on the North Slope, where the land is adjacent to major units of the Columbia Basin Irrigation Project.

Most of the interest in new industrial, especially high tech, development lies in a wide swath between Richland and the 200 East area, northeast of highway 240. Except in the area between the City of Richland and the Washington Public Power Supply System nuclear plant, proposed development is away from the Columbia River. This recognises the likelihood that Congress will protect the river corridor, reflecting the habitat and wildlife values found there. But there are habitat and wildlife values in much of the area where future economic growth is proposed. One can overlay species habitat and vegetation maps on the area to see the extent of potential conflict.

One concrete proposal, the Laser Interferometer Gravitational-Wave Observatory (LIGO), lies in an area of recovering shrub steppe habitat, damaged by wildfire in 1984. The LIGO project may in fact minimise land use conflict, because it will not greatly disrupt surface character, but will require a significant undeveloped area as a buffer. The proposed Superconducting Magnetic Energy Storage project, proposed for a site southeast of Gable Mountain, may pose greater conflicts.

In dealing with its site cleanup mission, the USDOE proposed a major landfill for contaminated soils to the south of the 200 west area. Partly on the basis of the Future Site Uses report, the proposal was moved away from the relatively undisturbed shrub steppe areas and closer to the southern boundaries of the 200 areas, where the land surface has been disturbed. But current planning documents still show a broad area, one where many habitat values exist, as potential waste management areas.

Considering again the area north and west of Richland, where future research, development, and engineering projects are anticipated, it should be noted that much of this area is underlain by nitrate and tritium-contaminated groundwater. Whatever activities may be developed here will be limited in their ability to use groundwater or to discharge surface water to the ground.

7.6   Conclusion

The foregoing illustrations indicate the nature of the conflicts that may be expected as Hanford moves from its historic mission to other uses. These conflicts have been analysed and presented using geographically coded information. As federal, state and local officials build good data bases, and as the public reviews and comes to understand them, they will be extremely useful in helping to find solutions to these apparent conflicts.

GIS data may, for instance, help these different interests weave new engineering complexes into a physical development plan that preserves major shrub steppe habitat and wildlife nesting areas. The use of GIS will help all parties define what the broad principles, including 'do no more harm' or 'protect the Columbia River', mean in specific cases.

In Washington State, our vision of Hanford is not just as a huge contaminated site. Hanford offers many opportunities, on a large scale, for new economically and ecologically productive land uses. Good geographic information and broad public commitment will both be essential if these opportunities are to be realised.

Note

1   Joy Denkers and Max Power are working at the Washington Department of Ecology, Nuclear Waste Program.

References

Hanford Future Site Uses Working Group (1992) *The Future for Hanford: Uses and cleanup, final report*, Richland, WA.
Power, M., et al. (1991) 'Future Site Use and Cleanup Strategy Alternatives: The Hanford Approach', paper presented at Environmental Remediation '91, Pasco, WA.
U.S. Department of Energy, Richland Operations Office (1993) *Hanford Site Development Plan*, USDOE, Richland, WA.
Washington Department of Ecology (1993) Hanford Land Transfer, Washington State Government, Lacey, WA.

Part B
Answers to Negative Environmental Spillovers in the Urban Area

Chapter 8

Risk-Based Urban Environmental Planning: The Seattle Experience

S. Nicholas[1]

8.1 Introduction

What is Comparative Risk Assessment? Simply put, comparative risk assessment (CRA) uses the conceptual and methodological framework of environmental risk assessment to compare the human health, ecological, and quality-of-life impacts of different environmental threats in a certain geographic area. It brings together the best available information about relative risks to help policy-makers, program managers, and citizens answer the question: 'Of all the environmental challenges facing our city, which are the most urgent?'

Comparative risk assessment is a key component of risk-based environmental planning, which uses comparative risk assessment to guide environmental priority-setting, strategy development, resource allocation, and environmental monitoring. Specifically, risk-based environmental planning uses information about risk – that is, information about the actual or potential harm caused by environmental threats – to structure and inform the environmental decision-making process in a community. The remainder of this chapter will focus on the larger concept of risk-based planning, including but not limited to the comparative risk component.

8.2 What is Risk-Based Planning?

Risk-based environmental planning has two major distinctive features – two characteristics that make it different from, and better than, the status quo. First, it is more information-driven. It encourages decision-makers to move beyond speculation, intuition, and media accounts of environmental threats, and to look hard at available data. What, if anything, do we actually know about these threats, the pollutants involved, emissions, concentrations, exposures, and effects?

Second, risk-based environmental planning, when done right, is more inclusive. Environmental decision-making is too often considered the exclusive bailiwick of the government and 'the experts'. But different people define and perceive risk differently, and there are usually large gaps in the available information that must be filled with human judgement and common sense. In addition, only the most broadly supported plans can hope to be effectively implemented. Recognising these realities, risk-based planning processes involve reaching out to different sectors of the community – to people from different parts of the community, with different backgrounds, different areas of expertise, and different philosophies.

In risk-based planning, the risk information augments, but does not supplant, human judgement. Risk-based planning is not an attempt to dehumanise the environmental decision-making process, as some critics have argued. On the contrary, if designed well, it rehumanises what has evolved into a highly bureaucratic and closed process, by involving all of the key stakeholders in the environmental decision-making of a community, and by providing a forum within which those stakeholders – typically people with similar goals but different views of the world – can talk, argue, educate, be educated, and, ultimately, locate common ground, agree on a shared vision for the future, and lay a foundation for working together toward that vision over time.

In short, risk-based environmental planning is not a substitute for the political process; it merely helps to structure it, and to infuse it with important information that historically has been overlooked or undervalued in the process. There are three key assumptions:

- *We need to set priorities*. Risk-based planning is predicated on the assumption that, because resources are limited, we do have to set priorities to ensure that we focus on the areas of greatest potential benefit. In the words of former EPA Administrator Lee Thomas: 'When everything is a priority, nothing is a priority.'
- *More is merrier*. A second key assumption underlying risk-based planning – and one that often goes unrecognised – is that a more open and participatory decision-making process is stronger (albeit messier and more difficult) than a closed one. Risk-based planning done well is not the sole responsibility of the 'technical experts'; it meaningfully involves policy-makers and 'plain people', as well.
- *Information is good*. Risk-based planning processes assume that more information about environmental problems – particularly information about the relative risks associated with those problems – is a good thing, and that failing to gather, study, and interpret this information can lead to bad decisions.

There are three key caveats:

- *Comparative risk assessment is a means, not an end*. Standing alone, the rankings of environmental threats generated in comparative risk projects, and the considerable technical work upon which they are based, are meaningless. What matters is what we do with that information. What actions do we take? In short, how do we use the process and the information it generates to improve the environmental quality and overall sustainability of our communities – that is, the overall quality of life of the people who live, work, and play in those communities, today and into the future?
- *Comparative risk assessment is a useful tool for environmental decision-*

making, but it is not the only tool. The history of environmental management in this country is marked by a frequent failure to consider available (albeit imperfect) information about relative risk. And, we are paying the price for that in some wasteful, narrowly focused environmental programs that are not focused on high-risk threats and/or do not address root causes of environmental ills and provide holistic solutions. Comparative risk assessment already has improved this situation dramatically in its relatively short history. But the risk information generated by comparative risk projects is only one factor in the equation of environmental priority-setting. Other important factors include public perception of the risk, and the feasibility (fiscal as well as technical) of reducing the risk. Most important is the collective judgement of the stakeholders, including both the people responsible for making environmental policy decisions, and the people who will be affected by them. Science, no matter how finely honed, will never be a substitute for human judgement.

- *Comparative risk assessment is an art, not a science.* While comparative risk assessment borrows heavily from its more scientific predecessor – quantitative human health risk assessment – it is not itself a science. In fact, it gets into trouble when it pretends to be – when it does not clearly communicate the considerable uncertainties inherent in the methodology, when it is not honest about the extent to which human values and other qualitative factors are incorporated into the process, in short, when it creates a veneer of objectivity and scientific certainty that is easily recognised as just that – a veneer. Comparative risk projects have emerged as a powerful tool for environmental management in this country and abroad. But this power emanates not from the numbers, but from the hard-fought consensus about *what those numbers mean for priorities and actions* – a consensus typically achieved by a broad-based, committed, well-respected group of policy-makers and citizens who live, work, and play in the community in question.

8.3   Risk-Based Planning: Where Did It Come From?

In about 10 years, comparative risk assessment has evolved from an intriguing idea worked on by a few staff people at the Environmental Protection Agency's headquarters in Washington, D.C. to one of the predominant environmental decision-making tools of the 1990s.

The publication in 1987 of the national comparative risk report, titled 'Unfinished Business', was a watershed event. In this study, about 75 experts from the U.S. Environmental Protection Agency completed a comparative risk assessment of 35 environmental threats, and found some discrepancies between the urgency of some environmental threats (measured as risk), and the amount of time and money being spent to address them. Thus began a national debate on the questions: Are we focusing on the 'right' problems? And, if not, what should we do about it?

The U.S. EPA followed up on the 'Unfinished Business' study in a number of ways. They began more focused comparative risk pilot projects in three of their 10 regional offices (Seattle, Boston, and Philadelphia), and they commissioned their Science Advisory Board (SAB) – a distinguished group of government and non government scientists and other experts – to assess the 'Unfinished Business' report and make recommendations for improvements. The SAB's 1990 report, titled 'Reducing Risk: Setting Priorities and Strategies for Environmental Protection', was another key event in the evolution of comparative risk assessment. It encouraged the EPA to continue to develop and use the comparative risk model in setting priorities and making other environmental decisions. And it made several specific suggestions for improving the process.

By the end of 1994, all 10 EPA regions had completed comparative risk projects, as well as several states and one city (Seattle). Comparative risk projects also were underway in several additional states, and in a handful of cities in the U.S. (e.g., Houston, Atlanta, Columbus, and the Cleveland area) and abroad (Bangkok, Quito, Cairo, and in various communities in Russia, Hungary, and Bulgaria). In addition, two major Western development agencies, notably the World Bank and the U.S. Agency for International Development, started using the risk-based planning model to help make lending decisions, and to guide client countries seeking to develop stronger environmental management programs.

In addition, two resource centres were established – one in Vermont and one in Colorado – to provide an information clearinghouse, co-ordination, and technical support to those undertaking regional, state, and local projects. Finally, the comparative risk-based approach was being discussed extensively by the U.S. Congress, as a separate issue as well as in the context of efforts to reauthorize major environmental statues such as the Clean Water Act, the Superfund Law.

8.4   Risk-Based Planning: Why Do It?

Risk-based planning processes are a lot of work. Why do them? There are a number of reasons why it makes sense. The major reasons are listed and described below.

Setting Priorities

The need to set clear priorities seems especially intense in cities. Increasingly, the responsibility for environmental problem-solving is falling on the shoulders of local governments, while state and federal funding have failed to keep pace (helping to spawn the 'unfunded mandates' debate). In addition, the competition for resources across policy areas (e.g. environment, housing, economic development, transportation, public safety, human services, etc.) seems more visible and pronounced at the local level: Will we invest in a swatch of open space downtown, or a new shelter for the homeless? These and other factors amplify the need for cities to think strategically about the environmental challenges facing their communities, and city government's role in meeting those challenges. The risk-based planning model helps cities do that.

Avoiding the Tunnel

Governments at all levels in the U.S. suffer from 'environmental tunnel vision', a tendency to focus on one problem or environmental medium (i.e. air, water, land) at a time. This is partly due to the human tendency to simplify a complex reality by compartmentalising and specialising. But it is largely a political and institutional phenomenon as well. We have written somewhat myopic laws and regulations to solve environmental problems, and we have created somewhat myopic institutions to make sense of, and implement, those laws and regulations. The air division deals with the Clean Air Act, the water division addresses the Clean Water Act, and so forth. This has made it difficult to take a holistic, co-ordinated, integrated approach to environmental management. The comparative risk approach addresses this problem, as well.

Improving Cohesion

The problem of fragmentation is especially pronounced in many cities. In Seattle, for example, some 15 different agencies in city government have environmental management responsibilities of some kind. Risk-based planning projects provide a mechanism for bringing these agencies together and improving co-ordination, communication, and camaraderie among them.

Clarifying Mission

The fragmentation in cities is exacerbated by confusion about mission. Environmental stewardship still is a relatively new addition to the urban agenda. It was not one of the city's primary reasons for being, such as, for example, filling potholes and providing public safety. Rather, environmental protection has been added to the mission of the city gradually over time, driven in part by a combination of public demands and regulatory requirements, and usually added to the work load of an existing department whose historical mission was closest to the new one. In Seattle, for example, the Engineering Department, originally set up to build and maintain the street system, now is the home for the city's bicycle and pedestrian program and street tree planting program. Few cities have an environmental department, per se. Consequently, in most cities the environmental mission is not as clear as it is for, say, public safety or transportation. Risk-based planning processes help clarify the environmental mission of the City, and the mission, roles, and responsibilities of each of its departments involved in environmental management activities.

8.5 Special Challenges, Special Rewards

Some of the special challenges to making risk-based planning work at the local level are mentioned above: the city's confusion over its environmental mission, a fragmented environmental management structure; a heightened competition for increasingly limited resources. A fourth special challenge is lack of local control over local environmental solutions expenditures. For example, many environmental problems are regional problems óver which the city has limited control. Transportation-related air pollution is a good example. This argues for increased regionalism in environmental management (see discussion in 'Challenges for the Future' section below).

In addition, a large percentage of the funding that cities spend on environmental programs is in response to state or federal environmental regulations and requirements. Sometimes, these national requirements are not appropriately tailored to local needs and priorities, and they diminish local government's flexibility in environmental problem-solving. (This goes a long way toward explaining why many cities strongly support federal legislation prohibiting so-called 'unfunded mandates'.) But along with these special challenges, there are special rewards, as well – comparative advantages. These include the following:

- *Proximity to the citizenry* Local governments are closer to the people. This is certainly a double-edged sword, but it generally bodes well for affecting environmental change, since many of the key environmental challenges of the present and future involve individual behavioural change as much as (if not more than) they involve technological change.
- *Better numbers* Where local environmental data exist, they typically are more accurate than state and federal data, since they are focused on the local situation, rather than extrapolated or interpolated from some broader database or geographic area.
- *Control over what counts* While cities currently lack control over what they do and what they spend in some big-ticket areas (e.g. compliance with the Safe Drinking Water Act) they have more control in some policy areas that are emerging as important factors in generating environmental risk, most notably land-use and transportation decisions. Cities also have control over their own corporate-like behaviour, as major landowners, employers, and service-providers. On the one hand, these activities make cities potentially major sources of environmental problems. But they also provide cities with unique and numerous opportunities to lead-by-example, to educate (and be educated), to use pricing policies to create market incentives for environmentally responsible choices, and to focus resources toward the development more environmentally friendly alternatives to their citizens.
- *Smaller playing field* Compared to states and the country, cities are smaller geographically, with fewer variations in, for example, demography, topography, and meteorology. This makes environmental risk analysis easier to do well.
- *Less bureaucracy* City bureaucracies typically are smaller and more manageable than those encountered in state and federal governments. In addition, the nexus between planning and budgeting, and the relationship between the executive and legislative branches of government, typically are tighter at the local government level. These differences make change easier at the local level – in theory, at least.

All of these factors bode well for risk-based planning at the local level.

8.6   Risk-Based Urban Environmental Planning: Seattle's Experiment

Seattle Mayor Norm Rice initiated a risk-based environmental planning process – called the Seattle Environmental Priorities Project – in 1990, shortly after his election. A few months later, in describing the project to the Policy Advisory Committee he appointed to guide it, Mayor Rice called it a way to change the way in which city government manages environmental issues, without re-organising the structure of the government itself (i.e., by creating a new office for environmental affairs). In particular, the Mayor cited his interests in taking a more integrated, holistic, co-ordinated approach to environmental problems and solutions in Seattle: What are the problems? Which are the priorities? What is the best way to address those priorities? What is City government's role?

That is what the risk-based planning process has been for Seattle – a 'middle path' of urban environmental management, between the status quo (which most people agreed was not working) and wholesale re-organisation (which some people preferred, but others viewed as too difficult to accomplish financially, organisationally, and politically).

The overriding goal of the project was to improve environmental quality and quality of life for the people who live, work, and play in Seattle. In pursuit of that goal, the project sought to develop and implement an integrated environmental action plan for the city, laying out a shared vision of the future, identifying priorities, and offering holistic, workable strategies for addressing those problems. To do that, the Mayor's Office, staffed by the environmental section of the City's Planning Department, designed and put in place an organisational structure (see Figure 8.1) and a four step process (see Figure 8.2). The organisation of the project, and each of the four steps, are described briefly below.

8.7   Organisational Structure

As shown on Figure 8.1, the Mayor appointed a 40-person Policy Advisory Committee to guide the Environmental Priorities Project, with the Mayor himself as chair. (This type of visible commitment and support from the highest elected official in the city was a key – perhaps the key – to the success of the project.) This committee consisted of senior-level environmental managers from city, regional, state, and federal government, business leaders, environmental activists, tribal representatives, academics, and neighbourhood representatives. The committee's most distinguishing feature was that it included both policy-makers and 'regular people'. Rather than creating a separate steering committee of government representatives and a separate public advisory committee of non-government representatives, the city made a conscious decision to create just one group consisting of representatives from both inside and outside of government. This was done in part to ease the work load. But mostly it was done to get the different perspectives around the same table, to talk to (rather than about) each other, agree on a vision for the future, and forge a consensus on environmental problems, priorities, and policy and program responses.

Figure 8.1   Environmental priorities project: organisational chart

The Mayor also appointed a 35-member Technical Advisory Committee (TAC) to lay the technical foundation for the process. This committee consisted mostly of environmental professionals working for city government, with a smattering of experts from regional, state, and federal government, academia, and neighbourhood groups. The TAC divided into four working teams (air, water, land, and cross-media) to get work done. Some projects have chosen to organise the technical work, and the committee that performs it, around types of risk (e.g., a human health risk team, an ecological risk team, and a quality-of-life team). This has the advantage of making the comparison of the problems along any one of the risk axes easier. The Seattle approach makes those particular comparisons difficult, if not impossible.

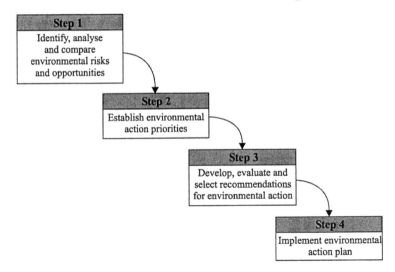

Figure 8.2 Seattle environmental priorities project: the process

However, it allows the technical team to take a more integrated look at each of the problems, and it simplifies the data-gathering activities considerably.

8.8 Identifying, Analysing, and Ranking the Issues

The TAC spent close to a year identifying the list of environmental problems to be considered and studied by the project (see Table 8.1), and analysing and ranking those problems. In conducting the comparative risk assessment of environmental problems, the TAC focused on three types of risk: human health risk (adverse impacts on human health, including both cancer and non-cancer); ecological risk (adverse environmental impacts, including impacts on both individual species of plants and animals and whole ecosystems); and quality-of-life risk (adverse impacts on the human condition, including impacts on aesthetics, recreational opportunities and economic opportunities). After sharing the results of their analyses, and discussing them at length, the TAC ranked the issues within each of the four areas, based on a combination of the relative risk information and their best professional judgement. The TAC's risk ranking is shown in Table 8.2.

Table 8.1   Seattle Environmental Priorities Project: Issues evaluated

Environmental Medium	Environmental Issue
Air	• Air Pollution from Motor Vehicles • Air Pollution from wood Burning • Environmental Tobacco Smoke • Other Indoor Air Pollution • Noise Pollution • Fugitive Dust from Industrial Areas and Unpaved Roads • Non-ionising Radiation and Electromagnetic Fields • Air Pollution from Gasoline Stations • Air Pollution from Industrial Facilities • Air Pollution from Yard Burning • Other 'Non-point' Sources of Air Pollution • Emissions from Centralia Power Plant
Water	• Water Pollution from Combined Sewer Overflows • Water Pollution from Stormwater Runoff • Contaminated Sediments • Direct Wastewater Discharges from Industrial Facilities • Ground Water Contamination • Accidental Spills into Lakes, Streams, and Puget Sound • Discharges from Commercial Vessels, Pleasure Craft and Liveaboards • Over-harvesting of Fish and Shellfish • Marine Plastic Debris • Drinking Water Contamination • Discharges from Wastewater Treatment Plants • Impacts from Hydroelectric Power Supply • Impacts from Public Water supply • Flooding and Other Impacts from Stormwater Runoff
Land	• Loss and Degradation of Greenbelts and Natural Areas • Loss and Degradation of Riparian (e.g., Creek) Corridors • Loss and Degradation of the Urban Forest • Loss and Degradation of Wetlands • Degradation of Parks
Cross-media	• City's Use of Hazardous Materials • Household Use of Hazardous Materials • Use of Hazardous Materials by Business and Industry • Leaking Storage Tanks • Abandoned Municipal Landfills • Inactive Hazardous Waste Sites • Land Application of Sludge

Table 8.2   Team rankings: relative risk[2]

	High Risk	Medium-High Risk	Medium Risk	Low Risk
Air Team	Transportation sources of air pollution Wood burning Environmental Tobacco Smoke (ETS)	Other indoor air pollution Industrial point sources	Gas stations Fugitive dust Centralia power station	Noise pollution Yard burning Other non-point sources
Water Team	Contaminated sediments Combined Sewer Overflows (CSOs) Stormwater discharges	Direct industrial discharges Accidental spills to surface waters	Impacts from water supply Impacts from hydropower supply	Boater discharges Other threats to the marine environment Ground water contamination Drinking water contamination Stormwater quantity
Land Team	Loss and degradation of greenbelts and natural areas Loss and degradation of riparian corridors	Loss and degradation of street trees and privately-owned trees	Distribution of and access to open space Degradation of parks	Loss and degradation of wetlands
Cross-Media Team	Environmental lead Toxics in house dust Use of hazardous materials by business and industries		City use of hazardous materials Household use of hazardous materials Land application of sludge Inactive hazardous waste sites	Storage tanks Abandoned municipal landfills

8.9   Setting Priorities

The Policy Advisory Committee (PAC) was charged with translating the TAC's work into a set of environmental priorities for the city. This part of the process, which took approximately six months, began with an intense series of meetings during which the PAC was briefed on the comparative risk analysis by members of the TAC and project staff. After this information exchange, often marked by heated debates about uncertainties and interpretation of data – the Policy Advisory Committee went through a ranking process of its own. The PAC selected its preliminary priorities based on four major criteria:

1   the vision of Seattle's environmental future that they developed at the start of the process;
2   the risk information provided by the TAC (and the PAC's interpretation of that data and analysis);
3   information about the city's ability to control or manage the risks; and
4   their collective judgement.

Citizens commented on the PAC's preliminary rankings in a series of four public workshops, held in different quadrants of the city. The PAC then met again several times to incorporate the public comments into their final rankings. Key differences between the TAC's ranking and that of the PAC are shown in Table 8.3.

8.10   Developing an Action Agenda

A new risk management committee was created to support the action plan development process. This committee consisted of selected members of the PAC and TAC, as well as some 'new blood' from both inside and outside city government, people with special expertise in the priority areas selected by the PAC. This committee organised into working teams corresponding to the 7 priority areas shown in the right column of Table 8.3, plus three cross-cutting management priorities (increasing pollution prevention, waste reduction, recycling, and conservation; improving environmental education and community outreach; and improving environmental management and co-ordination).

Each of these 10 teams was responsible for the following tasks: 1) brainstorming a list of potential action strategies for their issue (starting with lists of suggestions generated at the aforementioned public workshops); 2) screening those ideas on the basis of cost-benefit criteria and best professional judgement; 3) doing some limited analysis of the costs and benefits of the remaining ideas; and 4) making recommendations to the PAC about which action ideas ought to appear in the action agenda. The PAC reviewed these recommendations, and submitted a recommended environmental action agenda to the Mayor for his review and approval. This part of the process took approximately six months.

The Mayor's Recommended Environmental Action Agenda, submitted to the

City Council and released publicly in September of 1992, does the following: 1) introduces some guiding principles for environmental management by the City; 2) identifies the City's environmental policy priorities for the next 3-5 years; 3) offers a menu of about 150 recommendations for addressing the priority issues; and 4) proposes an implementation regime.

Table 8.3   From risk to policy

Risk Rankings	Policy Priorities
Air: • Transportation • Wood Burning • Tobacco Smoke	• Environmental Risks from Transportation Sources • Air Pollution from Wood-burning
Water: • Contaminated Sediments • Combined Sewer Overflows • Stormwater Discharges	• Water pollution from combined Sewer Overflows • Water Pollution from Stormwater Discharges
Land: • Loss of Greenbelts • Degradation of Riparian Corridors	• Loss of Open Space, the Urban Forest, and Habitat Areas
Cross-media: • Environmental Lead • Toxics in House Dust • Use of Hazardous Materials by Business and Industry	• Noise Pollution • Indoor Environmental Pollution • Global Climate Change

8.11   Implementing the Action Agenda

The Mayor's Recommended Environmental Action Agenda was adopted unanimously by the Seattle City Council in October 1992. In addition to endorsing the guiding principles of environmental management and the environmental priorities, and committing to further exploration of the 150 action recommendations, the Council's resolution put in place a mechanism for implementing the agenda. Specifically, the Council required the implementation of several actions as 'first steps', and it directed the City's Planning Department (now the Office of Management and Planning) to track implementation progress, and to report on that progress, to the Council and to the public, every two years.

The implementation process, like the process by which the agenda itself was developed – is a collaborative, interdepartmental effort co-ordinated by the City's Office of Management and Planning (formerly the Planning Department). The primary responsibility for implementing most of the action recommendations rests with one or more of the 15 agencies who do environmental management. The Office of Management and Planning co-ordinates, facilitates, and monitors the implementation of those recommendations, works to incorporate implementation needs into the City's budget process, and implements some action recommendations directly.

8.12   Measuring Progress

As mentioned earlier, the city closely monitors implementation progress, and has reported on the status of implementation in each of the last few years. In addition, the City is in the process of developing broad quality-of-life indicators and more program-based performance measures (including, but not limited to, measures of environmental quality and effectiveness). These – along with the 'Indicators of Sustainable Community' developed by a local non-profit group called Sustainable Seattle – will be used to measure the City's progress in implementing not just its Environmental Action Agenda, but a host of other plans, policies, and programs as well.

8.13   Success Stories

Has Seattle's risk-based planning process, which has been in place since 1992, been a 'success'? The jury is still out. But the progress to date has been impressive. The Vermont-based Northeast Center for Comparative Risk in 1994 completed an independent evaluation of the Seattle Priorities Project during which they interviewed more than 20 project participants. Their report concludes that 'Seattle generated more activity in its risk management phase than any other comparative risk project that we have reviewed'. A sampling of these activities is provided below.

- *More action* The City's Environmental Action Agenda has increased the City's net environmental activity. Environmental actions spawned by the Environmental Action Agenda, and the process by which it was developed, fall into the following four categories:

  1 *New initiatives* Several new programs or projects were developed, especially where gaps were discovered between the relative urgency of a problem and the City's existing response to it. A good example of a new initiative in Seattle is the Master Home Environmentalist Program - a response to indoor environmental risks, one of the risk-based priorities identified in the Environmental Action Agenda. The City joined with adjacent jurisdictions and local non-profit organisations to launch this effort, in which volunteers are trained to understand indoor environmental risks (e.g., exposure to household hazardous wastes), and to help households identify and reduce those risks.

  2 *Modifications to existing initiatives* In some cases, existing programs were modified based on the findings of the comparative risk process. For example, the City's noise pollution control program, which had

been significantly curbed the previous year, was resurrected and strengthened after noise pollution was identified in the Environmental Action Agenda as one of the high priority issues.

3   *Changes in corporate culture* An ethic of environmental stewardship is much more prevalent in the City's corporate culture now than it was, when the Environmental Priorities Project began. The City's overall environmental mission – that is, the City's roles and responsibilities vis-a-vis environmental preservation – has been clarified. And environmental issues are more readily incorporated into business-type decisions by the city, such as what sort of cars to purchase for the City fleet. One example is a new Municipal Resource Conservation Project, aimed at reducing the City's use of water and energy resources through retrofits and other conservation measures.

4   *Influence over other policy areas* The risk-based planning process in Seattle, and the Environmental Action Agenda that resulted, are having major impacts on other planning efforts in and around the city. The best example of this is the City's lately adopted growth management plan. This plan was designed in large part to address the two highest priority risk issues identified in the comparative risk assessment: environmental risks from transportation sources (mostly air pollution, but also water pollution and noise pollution); and loss of open space. The Action Agenda also has helped to shape the City's thinking about, and involvement in, a regional transit project that is contemplating major changes to the region-wide public transportation system.

- *Better co-ordination* Co-ordination among City departments responsible for environmental management has improved significantly. The network of environmental professionals established by the process still exists, even though the committees themselves have long since been disbanded. In addition, a regular mechanism has been established for bringing different departments together to tackle and solve common problems – problems that can only be solved interdepartmentally, such as the degradation of the urban forest in Seattle.

- *Better communication* Communication about environmental issues has improved as well, both inside City government and between the City and its citizenry. The Environmental Action Agenda provides a useful framework for talking about environmental issues and the City's response to them.

- *Better budgets* The City's Environmental Action Agenda has helped guide budget decisions in iterations of the budget cycle, by steering resources toward high-priority environmental initiatives. In fact, environmental management now is one of the few policy areas in which a strategic plan is in place against which rational resource allocation decisions can be made.

- *Heightened awareness* The risk-based planning process has heightened awareness of environmental issues among elected officials and throughout City government. In particular, the project served to sensitise several City departments to existing environmental management responsibilities, as well as new opportunities.

8.14   Challenges for the Future

One of the remarkable features of comparative risk assessment and risk-based planning is its adaptability. This is the key to its success and survival so far: it is an intuitively appealing approach that is flexible enough to adapt to changing circumstances and needs as weaknesses in the model are discovered, and as the model is applied in different geographic areas and different political contexts. Some of the key areas in which modifications may be necessary include the following:

- *Sustainability* Sustainability refers to the long-term environmental, economic, and social health of a community. Its two most distinguishing features are its holistic focus (i.e., its simultaneous focus on economic, environmental, and socio-cultural endpoints) and its long-term perspective (i.e., effects on future generations). In the wake of the 1987 United Nations World Commission on Environment and Development report titled 'Our Common Future', and the Earth Summit in Rio in 1992, sustainability is emerging as a major organising principle and policy-making parameter in cities around the country and throughout the world.

Risk-based planning must evolve into a tool that is useful and relevant to this movement. As an interim step, risk-based planning projects should include 'sustainability' as an analytic and priority-setting criterion: What are the relative impacts of the environmental threats being compared on future generations, and on economic and social (as well as environmental) health? Ultimately, however, project managers might consider applying the comparative risk approach to the larger sustainability context, that is, designing a process that systematically compares and sets priorities among threats to the overall sustainability of the community, not just environmental quality and public health. This will dramatically expand the scope of the projects, and quickly outstrip the capabilities of existing analytic methodologies. But this more holistic approach is the wave of the future. It may be time to begin testing the comparative risk model's ability to stay afloat.

- *Future focus* Risk-based planning projects run the risk (pun intended) of focusing primarily, if not exclusively, on past and present risks, i.e., the 'devils we know'. While many of the projects already completed have attempted to consider and analyse the trends of these risks (i.e., are they decreasing, increasing, or staying about the same?), few if any have tried to address the question: 'What are some of the environmental risks that may occur in the future, based on existing trends, experience in other communities, or other available information?' This will be a difficult but necessary step for future project managers to take.

- *Equity* Risk-based planning projects have failed to adequately address the issue of environmental justice: Are the risks to certain segments of the population (e.g., low-income people, people of colour) higher than they are for the community as a whole? If so, why, and what should be done about it? Comparative risk projects typically focus on sorting out environmental priorities for the community as a whole; they therefore risk overlooking, and even exacerbating, the problem of environmental justice. Future projects should consider this issue by building environmental justice concerns into both the comparative risk analysis (e.g., geographic-based analysis, analysis of correlations between environmental and demographic variables, etc.) and the development of the action plan or agenda (e.g., include a section on environmental justice-related actions).

- *Pollution prevention* Some critics of comparative risk and risk-based planning argue that it is a 'pollution justification' model; by focusing exclusively on identifying and ameliorating risk – particularly the most urgent risks – these projects can ignore root causes, and overlook promising opportunities to prevent risk in the first place. Risk-based planning project managers should consider different ways of avoiding this potential blind spot. Ideas for doing this include: including a 'root cause analysis' at the beginning of the process; including a section on across-the-board pollution prevention opportunities (i.e., regardless of the level of risk associated with the pollution being created) in the action plan; and using the results of comparative risk analyses to identify and focus future pollution prevention efforts.

- *Cumulative risk* Comparative risk analysis typically focuses on a set of discrete sources of risk – air pollution from cars, water pollution from storm water discharges, etc. There is some overlap between this issue and the environmental justice issue, discussed above. Geographic-based analysis, in addition to risk-specific analysis, may be helpful in gaining greater understanding of cumulative risks, and possible policy and program responses.

- *Regionalism* The regional nature of most environmental problems and solutions is becoming more and more obvious to policy-makers and citizens alike. The political boundaries that we created and within which we now operate have little if anything to do with the way in which environmental risks are generated, how they move around, and what needs to be done to reduce or eliminate them. Risk-based planning processes need to recognise this reality, and take on regional involvement and a regional focus – no easy task in the face of the turf-guarding tendencies that have developed over time among politicians and bureaucrats alike.

- *Indicators* The national movement (some would use the term 'craze') toward 'total quality management' and 'reinventing government', combined with voter demands for lower taxes, more flexible regulations, and greater accountability of government, have placed a spotlight on the need for better indicators of progress. How can we measure, and clearly communicate to both ourselves and our citizens, whether or not our environmental actions and expenditures are really making a difference? Risk-based planning projects and efforts to develop and monitor environmental indicators have been treated rather separately; closer ties between the two should be drawn. Comparative risk projects can be designed so that the results are useful in defining and/or measuring environmental indicators. Similarly, existing indicators can be useful in improving the relevance of comparative risk projects, for example in deciding which environmental threats to include, and how to analyse them.

Notes

1   Steve Nicholas is Senior Environmental Planner for the Seattle Office of Management and Planning. He is co-ordinator of the City's Environmental Priorities Project, the first urban risk-based environmental planning project, in the USA.

2   These rankings are based on each Team's collective judgement about the relative risks posed by the issues, including human health, ecological, and quality of life risks. Issues within cells are not in rank order.

Chapter 9

Towards an Integrated District Oriented Policy: A Policy for Urban Planning and the Environment in Amsterdam

E. Meijburg[1]

Summary

In 1994 the Amsterdam Departments of City Planning and Environmental Affairs have drawn up a policy document on urban planning and the environment in the city of Amsterdam. This document is intended to enable Amsterdam to make a maximum contribution to the development of a liveable and sustainable city. It contains a strategy and a number of tools to assist in finding inventive solutions to planning and environmental problems in the compact city. The strategy is based on the so-called *stolp* (bubble) method developed by the Institute for Environmental Studies of the Free University of Amsterdam.

9.1   An Integrated District Oriented Policy

Introduction

The 'environmental space' in a compact city such as Amsterdam is limited. City planning can only begin to contribute structurally to the realisation of a liveable and sustainable city when starting points for environmental protection are integrated into city planning policy. To this end, the Department of City Planning and the Department of Environmental Affairs of the city of Amsterdam have drawn up a policy document on city planning and the environment. This document provides a policy framework for district oriented, integrated city planning and environmental policy-making, at a municipal (city) level. This policy addresses three problems: the 'paradox of the compact city'; the integration of environmental and city planning; and the level at which solutions are found.

The Paradox of the Compact City

Amsterdam aims to improve living conditions in the city as well as contribute to sustainable development. In trying to serve both interests, one encounters problems relating to the so-called paradox of the compact city. The compact city policy, aimed at sustainability (concentration of functions, high density development, in or close to the city) has positive environmental effects at the macro level, but has negative environmental effects on the local level, since it increases the impacts upon living conditions in the city.

The Integration of Instruments

Both the city planning policy and the environmental protection policy are focused on the quality of the environment. However, both areas of policy use instruments that tend to hinder rather than support each other. Carefully co-ordinated instruments offer much better opportunities to create a desirable development.

Defining Environmental Problems at the City Level

Environmental effects occur on different levels, such as the city district or neighbourhood, the city as a whole, and the region. It is very important that all levels contribute to sustainable development. Each problem must be resolved at the most appropriate level, and measures taken at different levels may not be in conflict. This will prevent the situation where solving the problem in one district results in the emergence of the same problem in another district.

In The Netherlands at municipal level the Structure Plan is the driving force behind city planning. The Structure Plan outlines the desirable long term developments and provides a framework for drawing up and testing local zoning plans. Zoning plans are the most important instruments at the city district (local) level.

9.2   Integrated District Oriented Policy: The Possibilities

Amsterdam is working on the development of an integrated district oriented policy or 'stolp' policy. This 'stolp' method combines city planning and environmental analysis, fostering selection of those measures that will yield the best results both for the environment and the economy. The method uses a 'city stolp' based on the 'bubble concept' applied in the USA, to depict the current state of the environment in Amsterdam. It was developed by the Institute for Environmental Studies (Free University, Amsterdam). The idea is that within a defined area (the city stolp) certain trades-offs will be allowed between forms of pollution, as long as the total pollution is decreased. The city stolp shows the total degree of environmental pollution in the city at a given time.

The aim of the municipal environmental policy is to reduce the total degree of environmental pollution in Amsterdam. Currently, the degree of environmental pollution in Amsterdam does not exceed the absolute health threatening levels but does exceed the environmental target level, or 'negligible risk'. It is almost impossible to attain this target level without losing the environmental advantages of the compact city.

Therefore, the stolp policy works with an optimal level, which does guarantee a reduction of the degree of environmental pollution, but at the same time leaves room for new developments in the city. This optimum lies somewhere in between the current level of environmental pollution (the city stolp) and the target level for the city. In the first instance, the environmental policy must concentrate on reducing the city stolp towards the optimal level. In the long run, this optimal level, and therefore the city stolp, must be reduced further towards the target level.

It is not necessary that all parts of the city contribute equally to the reduction of the city stolp. In the integrated district oriented policy, this contribution depends on the characteristics, functions and spatial possibilities of this district. Measures must be geared to the specific characteristics and future prospects of the districts. Each district has its own sub stolp, so to speak, within the city stolp. Each district aims at optimalisation. When a spatial or environmental measure has virtually no result in one district (from an economic as well as from an environmental point of view), other measures must be taken to enhance the environmental quality of the district; or improvement in environmental quality must take place elsewhere (but within the general city stolp) by additional measures. This way it is possible to select those measures with the best environmental results for the city as a whole.

The Stolp Method

The stolp method pursues an integrated approach, aimed at combining several fields of policy and consideration of several forms of environmental pollution, including the ways they are related. It is a flexible method, in that it allows the exchange (trade off) of different forms of environmental pollution. It also provides insight into the costs and benefits of the measures and the motivation behind them.

The method is based on the idea that a substitution of environmental measures according to location, source and category of environmental pollution, may lead to better results than would a uniform standard. In the final version of this method, different forms of environmental pollution will be exchangeable. There are four exchange possibilities:

1   exchange of one form of environmental pollution between different parts of Amsterdam, for example, by allowing more air pollution in some districts than in others;
2   exchange of different forms of environmental pollution in Amsterdam, for example by allowing more noise, but less air pollution in a given district;
3   exchange of one form of environmental pollution between Amsterdam and the surrounding areas;
4   exchange of different forms of environmental pollution between Amsterdam and the surrounding areas.

Initially, the proportion of environmental pollution per source per category, of the total (100 per cent) environmental pollution, will be determined. Until now, the assessment of the degree of environmental pollution in Amsterdam has been based solely on available data dealing with noise, industrial risks, and air pollution.

For purposes of comparison, current levels of environmental pollution for each category were transformed into non-dimensional numbers (i.e. 'index'). In order to do this, different forms of environmental pollution are indexed, in other words, the degree of pollution was expressed in a number on a scale from zero to a hundred. Zero represents the legal target level (negligible risk or hindrance) and one hundred the limiting level (highest permissible risk or hindrance). Multiplication of the index by the number of people impacted by that form of pollution determines the size of the stolp in each district.

The translation of pollution levels into an index is done in such a way that these indexes can be combined. That way, the current degree of environmental pollution can be calculated per district (stolp). The city stolp shows the total degree of environmental pollution in Amsterdam, the city stolp being the sum of the various (sub) stolps in the total city. This is the total index.

Once the total degree of environmental pollution is known, the exchange or substitution possibilities can be assessed. When these calculations are made, those districts where the degree of pollution is relatively high and those where there is relatively little pollution will be identified. Then the environmental measures yielding the most positive environmental effects are studied. The decision as to where the environmental pollution must be reduced depends not only on the largest environmental improvement possible, but is a function of a location, the spatial objectives, the population, and the costs of the environmental measures. The indexes of the different stolps and the general index of the city stolp provide the basis for considering possibilities for exchange.

In determining the exchange possibilities, a complete city stolp method must include an index of the local living conditions. Until now, attempts to express liveability in a single figure have failed.

Finally, an assessment will be made of the cost effectiveness of projects to improve environmental quality. The best results will be obtained by those measures that yield the biggest environmental (or 'index') improvement and are least expensive. This will be linked to the function and population of the concerned location. It will be possible then to think of measures to reduce the environmental pollution in each district. This can be done, for example, by reducing the level of noise (and thus the index will drop), and the number of people annoyed.

Attention will also be focused on new ways of obtaining funds for environmental improvement. Then, a final assessment is made. This assessment depends on the financial and exchange possibilities and the environmental results of the various measures. When this method is applied, the measures having a positive effect on the total degree of environmental pollution in the city will be made apparent. When these measures are carried out, the present city stolp will shift downward to that city stolp which is most realistic at that moment. In due course, when new budgets and environmental measures become available, this stolp will be further reduced.

The city stolp survey will be an important step towards an integrated district oriented policy. This environmental survey clearly indicates the current state of affairs, and points out the gaps in data requiring further research. As yet, the method is not operational and needs further development. However, the philosophy behind the stolp method is at the basis of the strategy for the integrated district oriented policy.

9.3   Integrated District Oriented Policy: The strategy

Compared to the traditional Dutch environmental policy, integrated district oriented policy should consist of several new elements, including:

- a means of adding up of different forms of environmental pollution;
- district oriented standardisation;
- result analysis;
- exchange and compensation.

Adding Up of Environmental Pollution

Various forms of environmental pollution should be added up to one total expression of the environmental situation. Such an 'overall picture' is necessary to gain better insight into the actual degree of environmental pollution in Amsterdam.

District Oriented Standardisation

Most environmental standards now in use have been laid down nationally in The Netherlands. However, environmental effects of interventions differ for each district, and this strongly depends on the district's sensitivity to environmental pollution. This sensitivity in turn, depends on the objects and functions one wishes to protect in a given district. The standards required in a densely populated residential area differ from those required in an industrial or scenic area. To a certain extent, this has already been considered since, for example, standards for noise and risks are linked to residences, homes for the elderly, and schools. These are regarded as critical development. Scenic areas may require lower acceptable levels of pollution as well. In general, the environmental policy distinguishes target levels which are low, limiting levels which are higher, and health threatening levels which are higher still. Limiting levels (mostly laid down by law) may not be exceeded. These levels however do not exceed absolute health threatening levels. It is not allowed to deviate from (to pollute more than) target levels except in special circumstances. Generally, the actual degree of environmental pollution, especially in the compact city, is closer to the limiting level than to the target level.

Result Analysis

Hopefully in the future it will be possible to calculate the results of environmental measures, ensuring a well-balanced assessment of possibilities. According to national policy, the target levels of environmental pollution should be attained on all locations, but in most cases this demands a disproportionate effort and requires more money than provided by the available budget. Therefore, better insight into the environmental and economic results of suggested measures is necessary, thus allowing a more balanced assessment of which measures are most effective for a certain district and contribute most to a better environmental situation in the entire city.

Exchange and Compensation

In the future we expect it will be possible to exchange and compensate different forms of environmental pollution and improvement among districts. When it appears that measures in a given area either yield relatively little result or are too expensive (for example, the reduction of noise), other measures to benefit the environmental situation in the same area might be given consideration (such as the reduction of air pollution). One could also try to compensate for the negative situation at some location outside the area, by for example creating a 'quiet area' such as a park just outside of the area. This would facilitate improving the environmental situation for residents whereas the traditional environmental policy would not have changed anything. This may mean that a local increase in some forms of environmental pollution must be accepted, when other areas can be spared.

By linking these elements to the spatial characteristics of a given district – in this case the municipal agglomeration of Amsterdam – a strategy can be formulated as part of the integrated district oriented policy. This strategy includes the following steps:

1   assessment of spatial development that is most appropriate for a given district according to the location of the district within the structure of the city and the region, the functional characteristics and the expected environmental effects;
2   determining the existing and future environmental pollution in the district where developments take place;
3   defining possible measures to reduce environmental pollution in a given district;
4   selecting measures that yield the best environmental results for the city as a whole, also with regard to spatial specifications;
5   exploring the possibilities of exchange and compensation, if desirable from the perspective of environmental results.

9.4   Integrated District Oriented Instruments

In connection with systematic integration of environmental issues into city planning, several new instruments are being developed in Amsterdam. These are important for a more district oriented approach, and include analysis of the environmental structure of the city, developing an Environmental Matrix, and an Environmental Exploration.

Analysis of the Environmental Structure

The current environmental structure of the city has been analysed from an historical perspective. The city can be divided into areas on the grounds of specific spatial similarities such as, for example, the medieval city, the ring of canals, the 19th century belt, the docks, and the remaining residential and commercial areas.

The division of the city into spatial systems or weaves is based on six typical characteristics:

- size and scale, density and height of the buildings;
- usage;
- function (and changes in that field);
- degree and scale of merging of functions;
- quality of (urban) development and living conditions;
- cultural historical significance.

By linking an environmental evaluation to the typical qualities of a spatial system, a specific district oriented policy can be defined, indicating which qualities must be either protected, changed or supported.

Environmental Matrix Structure Plan

A survey executed by the Department of City Planning on the relation between certain spatial developments and the environmental effects has resulted in the Environmental Matrix for the Structure Plan. The Environmental Matrix functions primarily as an indicator, and offers a quick and simple assessment of environmental effects and urban developments.

Three environmental themes: 'environmental nuisance', 'green zones in the city', and 'mobility', determine the direction of development in parts of the city that is most appropriate from the perspective of the environment. These environmental themes were based on the following assumptions.

Environmental Limiting Conditions

Based on the current legal standards with regard to noise and external safety risks, sources of nuisance have been outlined within which critical functions (particularly housing) are adversely impacted. Major impacts which limit future development include noise nuisance caused by air traffic, noise nuisance caused by industrial activities, and external safety.

Green Zones in the City

A green zone structure has been established for the city requires protection. This green zone structure includes elements that are of fundamental importance for a coherent structure of the green zones in and outside the city. This green zone structure includes those elements that are or could possibly become important for the city in more than one way. The mainframe green structure is based on sub-themes:

- cultural historical values;
- ecological values;
- recreational values;
- spatial structure;
- regional coherence.

Mobility

The objective here is to discourage use of the car and to encourage use of public transport and bicycles. This follows on the philosophy of the Dutch district oriented policy: places that are easy to get to by public transport must be used intensively. A place is considered as easy to get to when over 75 per cent of the surrounding population can reach it within 45 minutes.

Based on these three aspects, future development possibilities that are most desirable from the point of view of the environment can be mapped out for each district. See Table 9.1 for an overview of the considering factors.

The Environmental Matrix, used at the Structure Plan level, tests the proposed spatial location for a development on environmental. The matrix includes three aspects:

- aspects of city planning that are relevant at city level;
- aspects of city planning that are connected to these environmental aspects;
- an assessment system providing insight into the environmental effects of a given proposal for city planning.

Table 9.1   Environmental matrix (partial)

Location within environmental contour	Location within mainframe green structure	Can you get there by public transport?	Strategy
No	No	Yes	Intensive use housing, office buildings, facilities
No	No	No	Housing
Yes	No	Yes	Intensive use: office buildings
Yes	No	No	Employment
Yes/No	Yes	No	Green zones
No	Yes	Yes	Intensive housing facilities, office buildings, green zones
Yes	Yes	Yes	Intensive employment, facilities, green zones

Aspects of City Planning that are Relevant at City Level

Aspects of city planning may vary per scale. In addition, not all aspects of city planning have consequences on every scale, nor can they all be solved by city planning policy. Therefore the following aspects are relevant at city level:

- noise and industrial risks (immision);
- the amount of nuisance that a planned function causes (emission);
- exhaustion of energy and raw materials caused by traffic;
- exhaustion of energy and raw materials from use of building materials;
- disturbance of 'quiet places';
- deterioration of scenic areas.

Aspects of City Planning Relating to Environmental Aspects

City planning relates to and can influence several environmental aspects of districts, including the characteristics of a district, such as sensitivity, accessibility resulting from the district policy, and location in relation to the urban area and population.

The Assessment System

Aspects of city planning can be linked to indicators which measure the severity of the environmental effects resulting from a proposed development. This is expressed in a score that depends, among other things, on the function (for example, housing, office buildings, nature). Based on this score, it can be determined which function is most appropriate for a given district from an environmental point of view.

The final assessment was based upon a multi-criteria analysis, so that in the future various environmental aspects can be expressed as certain values. An example of the environmental Matrix is given at Annex A. The Environmental Matrix can be used in four ways:

- when having to decide upon a purpose for a given district (which function is most appropriate for a certain location);
- in selecting a location for a single function (which location is most suitable for a given function);
- in optimising a function for a location (which measures are most suitable for improving the environmental situation);
- in determining priorities for certain developments (which location must be developed first, from the point of view of the environment).

With the Environmental Matrix it is possible to make an environmental assessment for the purposes of housing (including small-scale activities, local facilities, and local green zones), office buildings, companies, large-scale facilities and nature. The instrument needs tuning in order to be able to assess recreational purposes (such as allotment gardens and golf courses) and large-scale infrastructural projects.

The Environmental Matrix is not a legal testing instrument. It provides a quick and simple assessment of the environmental pros and cons of different developments in various fields. Thus, environmental aspects can be integrated as guiding principles into the decision process at an early stage. For example, based on the outcome of such assessments, it is possible to give advice and guidelines regarding city planning or the phasing of different projects.

Environmental Exploration

In order to map environmental pollution, the Amsterdam Department of Environmental Affairs has established the so-called Amsterdam Environmental Exploration. The objective is to systematically map the quality of the environment in the city as a whole and for different parts of the city. The Environmental Exploration attempts to describe the quality of the environment in the broadest sense of the word. The quality of the environment in Amsterdam is divided into three elements:

- the physical chemical quality of the environment, for example the contribution of Amsterdam to the national discharge of acid equivalents;
- the impact on the environment of production and consumption; the environmental situation is linked to functions such as housing, employment, traffic and nature, presenting the processed data per function for small units; for example for 'housing', the energy consumption is calculated per household unit;
- people's experience and appreciation of the environment.

The first Amsterdam Environmental Exploration, produced in 1994, contains data not all in the desired form, nor will they be sufficiently accurate. In addition to the assessment of the quality of the environment, the main outcome of this first Environmental Exploration is the identification of gaps in expertise. The follow-ups of the Environmental Exploration will fill these gaps and developments can be assessed.

Within the framework of the integrated policy, in addition to more information about noise, air pollution and industrial risks (these are the environmental aspects that have been considered in the stolp method) a systematic record of other forms of environmental pollution must be kept. Also, research must be done on how the inhabitants of Amsterdam define the notion of liveability. Then a parameter for the city stolp can be developed based on the information of the Environmental Exploration, which will allow a comparison of various forms of environmental pollution.

9.5   Conclusions

This chapter outlines how the Department of Environmental Affairs and the Department of City Planning of the city of Amsterdam are working together on new initiatives, with the aim of improving environmental quality through an integrated, district oriented policy. Important new elements of an integrated district oriented policy are:

- the result of environmental measurements should be compared to the demands of city planning;

- measures for different forms of environmental pollution should be added up;
- the use of a district oriented standardisation should be made possible;
- different forms of environmental pollution are exchanged and a local increase of environmental pollution should be compensated in other locations, in or in the vicinity of the city.

In addition, the compact city policy will continue, on a city level, with an emphasis on:

- concentrating and merging of functions at strategic locations;
- reducing the growth of car mobility by continuation of the location policy;
- prevention of situations that are unacceptable from the point of view of environmental quality;
- guaranteeing a high quality green structure for the city.

The city stolp will need further development and improvement in years to come. The Environmental Matrix also needs further elaboration. An important role has been reserved for the Environmental Exploration, which systematically collects the necessary environmental data on the city.

The main objective remains to reduce environmental pollution in the city. Although there is still work to be done before an integrated district oriented policy will be fully operational, the philosophy behind it will serve to shape the agendas and methodologies of city planning and environmental policy.

Note

1   Emerentia Meijburg is working at the Amsterdam Department of Environmental Affairs, Amsterdam, The Netherlands.

Annex 9.1   Location and pollution

	A Immission		B Emission		C Mobility		
multiplier	Mlp A = 2		Mlp B = 3		Mlp C = 3		
1 location within polluted or critical district	location within environmental contours (maps 1a till 1c: assess score per map): ho, critical mf mo, mf is, na	 -2 -1 0					
2 location within city weave							

3 accessibility by public transport					Accessibility (map 3a/3b)
					very easy:
					  is, na
					  ho
					  mo, mf
					
					easy: mo, mf
					  is, na
					  ho
					
					poor: mo, mf
					  ho
					  is,na
						PT
						
						-2
						-1
						0
						
						-1
						-1
						0
						
						-2
						-1
						0
							Bike
							
							-1
							0
							0
							
							-1
							0
							0
							
							-2
							-1
							0

4 degree of pollution			category of environmental pollution: industry business wo, ka, vz, na	 -2 -1 0			
5 space							
6 density/ intensity			number of those hindred, population within a radus of (400m)(map 2): ≥ 200 people/acre 120-200 p./acre ≤ 120 people/acre wo, ka, vz, na	 -2 -1 0 0			

ho   = housing (including local facilities, small-scale activities and city green)
mo   = main purpose offices
mf   = metropolitan facilities
is   = industrial sites
na   = nature

		D Exhaustion of energy and raw materials		E 'Quiet' space		F scenic areas	
multiplier		Mlp D = 3		Mlp E = 4		Mlp F = 4	
1	location within polluted or critical district			deterioration of park space in districts with high density population (map6) ho, mo, is, mf: deterioration of remaining parks: ho, mo, is, mf not applicable ho, mo, is, mf	-2 -1 0	location within regional ecological structure (map7): ho, mo, is, mf na location within valuable land, outside region.ciol structure: ho, mo, is, mf, na location outside valuable land: na ho, mo, is, mf	-2 0 -1 0 -1 0
2	location within city weave	location compared to urban area: separate from ula. ho, mo, mf, is na outskirts: ho, mo, mf, is na inside urban area: na ho, mo, mf, is	-2 0 -1 0 -1 0				
3	accessibility by public transport						
4	degree of pollution						
5	space			surface reduction existing park: more than 1 acre less than 1 acre	-1 0	surface reduction existing scenic areas (only wo, ka, be, vz): > 1 acre < 1 acre	-1 0
6	density/ intensity	urban environment (map 5): ho.envir. IV low-rise Idem, III+IV layered Idem, I + II office monofunct. Idem, merged business extensive Idem intensive nature	-2 -1 0 -1 0 -1 0 0			Legend housing, (incl. local facilities, small-scale trade, city green)	

Chapter 10

Improving Environmental Performance of Local Land Use Plans: An Experiment with Sustainable Urban Planning in Amsterdam

E. Timár[1]

Summary

In The Netherlands, the local land use plan is a tool for safeguarding local environmental quality. This is accomplished by three means: the regulation of noise and hazards, the enforcement of location policy, and environmental zoning around industries. Unfortunately, by law the local land use plan is not equipped to actively enhance sustainable development of an urban area. The City of Amsterdam therefore developed an experimental Environmental Performance System (EPS), which is plugged into the local land use plan of a recent project. The Environmental Performance System is quantified in an Environmental Performance Level. Building Permits are only granted after obtaining the required performance level. To obtain this level, some discretion is possible in selecting from a list of 17 measures that are meant to promote sustainable urban development.

10.1 Introduction

In The Netherlands, the local land use plan as used for several decades is a strong tool for safeguarding local environmental quality. Unfortunately, the power of the local land use plan is restricted to preventing of environmental nuisances (e.g. noise, hazard), and is not equipped to improve upon existing environmental quality in an urban area. National environmental (sectoral) laws do not compensate for this effectively. Furthermore, environmental laws and the methodology of local land use plans do not facilitate a region-oriented environmental policy.

Amsterdam, an historical city with all the complex environmental problems of a capital city, is developing a region-oriented environmental policy, which includes 'sustainability standards' that are not yet provided for in the sectoral environmental laws. In Amsterdam this approach has lately led to the presentation of two planning instruments, the stolp method discussed in another chapter in this volume, and the

Environmental Performance System (EPS), which is the focus of this chapter. Before discussing the Environmental Performance System, attention will be paid first to the environmental role of local land use plans in general. This will be followed by the presentation of the Amsterdam approach in developing the local land use plan, particularly the experiment with an Environmental Performance System. This experiment will be illustrated with the description of an urban planning project in Amsterdam; the IJ Bank project.

10.2   Local Urban Planning in The Netherlands

In The Netherlands the local land use plan is the most important urban planning-instrument for each of the about 500 municipalities. The local land use plan serves as a tool for development and modification of an area, but also as a means of maintaining the quality of existing areas. This plan is the operational framework for the day-to-day development decisions affecting surroundings in the neighbourhood. In The Netherlands local land use plans usually consist of the following mandatory parts:

1   planning maps of the area;
2   descriptions of the outlines of desired developments;
3   regulations which serve as limiting conditions to the developments;
4   a survey of the environmental nuisance of various categories of companies;
5   an acoustic report on the area.

The local land use plan determines what kind of and how many houses, shops, companies, sport accommodations, public services etc. are permitted and where they will be situated. Also the height of buildings and the location of parks, parking spaces, canals, railway lines and roads are laid down. This plan indicates whether companies are allowed in the area and whether they may be combined with houses. At this planning level, exposing people to environmental nuisance has immediate consequences and there is a very large demand for legal protection among citizens.

For several forms of environmental nuisance (for example soil pollution, sound nuisance and industrial hazard) statutory protection standards have been laid down. Therefore, the local land use plan provides the legal guidelines for urban development. The more creative, detailed aspects of design usually take place after the local land use plan is drawn up.

The integration of more environmental aspects into local land use plans has been promoted in The Netherlands since the nineties. The City of Amsterdam contributes to this innovation, forced by the growing complexity of environmental problems and the subsequent need for integration of environmental interests along with economic and social interests in urban planning.

10.3   Environmental Aspects in Local Land Use Plans

The Dutch environmental legislation makes use of environmental quality requirements, which have been translated into standards. These standards can describe the desirable quality of the environment under several circumstances (physically, chemically, biologically). The standards are legally binding. Unfortunately, the current legal possibilities to include these environmental quality requirements in local land use plans are only limited. It is, for example, impossible to describe a desirable ecological diversity of an area by means of a local land use plan, or to treat energy efficiency targets for a city district. It is only possible to include urban planning measures that reduce the environmental nuisance of companies and traffic infrastructure (so-called 'mitigating' measures). The strongest tool in this field is environmental zoning, which will be illustrated below.

Noise Nuisance and Hazards

Around major sources of noise (companies, traffic infrastructure) contours are laid down within which development of 'sensitive uses' is limited or in some cases even impossible (e.g. noise levels higher than 55 dB(A) in new urban developments). By 'sensitive uses', house construction is meant in particular, as well as some facilities such as nursing homes and hospitals. The same approach is followed in the case of industrial hazards, expressed as levels of individual risk to inhabitants.

Location Policy

On a national level, a location policy for offices and companies has been formulated in order to reduce unnecessary car traffic and to improve the liveability of the city (Ministerie van Verkeer en Waterstaat 1990). Several kinds of locations are distinguished:

- A-locations can be reached very easily by public transport and bicycle;
- B-locations can be reached easily by public transport, bicycle and car;
- C-locations can be reached very easily by car.

Local land use plans specify how many parking places can be created. At A-locations, relatively few parking places will be permitted. Companies which necessarily depend on accessibility by car are only allowed at C-locations. Companies which do not necessarily have to be easily accessible by car may be placed at A- or B-locations. Undesirable and unnecessary car traffic can be prevented by this policy in some places, and the use of public transport can be promoted.

Environmental Nuisance of Industries

The environmental nuisance of industries in urban areas is regulated in two ways; by local land use plans and by environmental permits for the companies. Permits are issued to industries under national environmental laws (especially the Environment Act 1994). These laws prescribe the use of certain environmental abatement technologies (e.g. dust filters, noise isolation etc.). The local land use plan specifies whether or not a certain type and size of manufacturing facility is allowed in the local area, and at what distance from other functions (housing, infrastructure, sports, shopping etc.). In short, the local land use plan prescribes the mandatory distance between industrial activity and other activities in the area. This methodology of environmental zoning has been developed by the national organisation of municipalities and is called the VNG-method (VNG 1992). For each sector of industry, this method indicates what minimum distance should be maintained from 'sensitive uses', based upon impacts on the surroundings in terms of noise, odour, hazards, pollution, traffic generation and visual appearance. This total impact results in 6 distance classes between 10 and 1,500 meters.

10.4   Area Oriented Environmental Policy in Amsterdam

National environmental laws and the methodology of local land use plans (including the environmental zoning methodology) do not facilitate a region-oriented environmental policy. Amsterdam is currently trying to develop a more region-oriented, or location-oriented environmental policy. This approach has already produced an Amsterdam method of environmental zoning and more recently fostered development of two planning instruments, the stolp method and the Environmental Performance System (EPS).

Amsterdam Method of Environmental Zoning

The VNG method of environmental zoning is difficult to bring into practice in a country like The Netherlands: an industrialised area with one of the highest population densities of the world and the absolute highest number of automobiles per square kilometre. This is even more the case for the City of Amsterdam, with its dense urban design. The VNG-method calculates, for example, that a restaurant should be located at least 30 meters from housing, and for a district heating unit (heat/power generation) that a distance of 100 meters should be maintained. In a city like Amsterdam, these requirements are infeasible and would lead to dramatic spatial separation of functions. A dynamic, lively city would change into a set of monoculture districts of houses, offices and catering establishments. Companies would necessarily move to industrial parks, generating much commuter traffic. Mixing companies and houses is generally regarded in Amsterdam as a contribution to the quality of the city. Therefore, the nuisance caused by certain functions has to be carefully balanced against their contribution to the viability of the city.

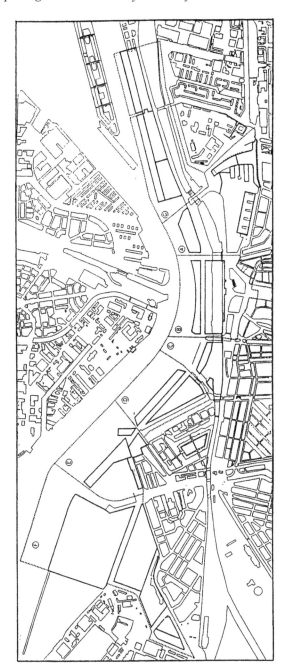

Figure 10.1   General design of the IJ Bank Project

In the early 1980s, the City of Amsterdam developed an alternative method of environmental zoning, in order to avoid adverse side effects of the zoning instrument. Manufacturing plants are grouped differently than the VNG method, into five categories of increasing environmental nuisance. This list is also included in the local land use plan as an appendix. Companies of category I and II are allowed in areas occupied by sensitive functions. In this case the zoning distance is zero meters. Companies of category III can be allowed near residential areas after individual assessment by the municipality. The category IV and V companies are directed to small industrial parks in the city, or to sites meant for heavy industrial use. This method is more applicable in Amsterdam than the VNG method, and supports city renewal without expelling half its historical functions. Urban functions like a restaurant, garage, district heating unit and small scale industries, can be mixed with housing without imposing distance requirements. In this way, a lively, sustainable city can be guaranteed, without exposing people to intolerable levels of environmental hazards and nuisance.

10.5   The Amsterdam Environmental Performance System

In The Netherlands, local land use plans are restricted to preventing environmental nuisances (e.g. noise, hazards), but are not designed to enhance sustainable development of an urban area. National environmental (sectoral) laws do shield people from intolerable levels of exposure to hazards (air and water quality standards, soil sanitation and so on), but do not direct the urban planning towards accomplishing sustainability in development projects. For this reason some municipalities in The Netherlands are conducting experiments with additional environmental requirements in the local land use plan (e.g. local land use plan 'Morra Park' Drachten 1993). Such additions are neither permitted nor expressly forbidden in the Spatial Planning Act. Amsterdam has taken this opportunity to experiment with a method of inserting elements of sustainability into the local land use plan of a large urban redevelopment in the central city, the IJ Bank project.

The IJ is a broad watercourse dividing Amsterdam into a northern and a southern part (see Figure 10.1). The IJ is connected to the North Sea through sluices. Over the last few centuries, the IJ Banks had an important function as an inland port with shipyards and industrial activities. However, some decades ago, the area fell into disuse and began to deteriorate. An ambitious plan is currently prepared to redevelop the southern bank, 280 hectares in size. In the next ten years, about 4,000 homes, a metro line, and hundreds of thousands of square meters of offices, manufacturing and other facilities will be built on a narrow strip between the railway and the watercourse. The intended high concentration of working, housing, public facilities and traffic on the IJ Bank makes assuring an optimal environmental quality there a challenge. Problems in the area are in particular traffic noise, soil pollution, bad water quality and lack of 'green' areas.

The specific situation on the IJ Bank also offers opportunities to achieve environmental successes such as reuse of non-functional buildings through remodelling, energy saving, soil clean-up, implementation of the compact-city policy and ecologically sound new building. This challenge has led to the development of an Environmental Performance System (EPS) as part of the local land use plan (Physical Planning Department of Amsterdam 1994). The EPS functions as a set of guidelines for the development of the IJ Bank, and is complementary to the defensive nature of the sectoral environmental laws. For the redevelopment of the IJ Bank the use of an Environmental Impact Assessment (EIA) has been deliberately avoided. Wide experience with the EIA suggested that this instrument, though useful for analysing the environmental impact of activities, is highly inappropriate as a means to guarantee any subsequent application of environmental considerations.

The EPS requires a minimal environmental performance from each area and building project within the boundaries of the local land use plan. The EPS employs a quantified Environmental Performance Level (EPL). Each building project on the IJ Bank has to realise an EPL of at least twenty points in order to receive building permits. To achieve this EPL, a choice from a list of measures is possible (see Annex 10.2). There are no restrictions on the choice of measures in each building project, as long as the minimal EPS is realised. Annex 10.2 illustrates how the EPS is applied in the case of the IJ Bank. An explanation of the EPL indicators is given in the appendix to this chapter (Annex 10.1).

Some citizens of Amsterdam expressed their doubt whether an EPL of 20 points was an acceptably high expectation. After some hypothetical tests of the EPS, it became clear that the EPL is rather ambitious and will demand substantial additional investments by the municipality and private investors.

In the EPS of the IJ Bank Project, requirements regarding ecologically sound building materials play an important role. Actually, such requirements do not really belong in a local land use plan, but rather in national or municipal building regulations. Unfortunately, the national government has forbidden municipalities to issue their own rules as regards the use of building materials. Consequently, a rapid increase of ecologically sound building has become difficult in The Netherlands. Therefore, in the EPS the attempt is made to bypass building acts and introduce legally binding building performance demands in urban planning.

10.6   Conclusions

Present provisions in national legislation concerning local land use planning, and the national laws regulating pollution by sector, do not facilitate a region-oriented environmental policy aimed at sustainability. Amsterdam therefore recently developed the Environmental Performance System (EPS), which is employed in developing and implementing local land use plans. As an experiment, the EPS has been applied in a major urban development project. The Environmental Department of Amsterdam in co-operation with the Urban Planning Department will undertake more research on this instrument in order to make the EPS applicable for all local land use plans of the city, including the suburban and rural parts. But before broad application of the EPS is considered, the IJ Bank experiment will have to be evaluated. Especially the following questions will have to be answered:

1   which adjustments should be made to the environmental indicators used in the EPS? Are they representative of all major aspects of 'sustainability' or should sustainable urban development be made operational with other indicators? Do the given scores for certain measures actually correspond with their relative environmental impact?;

2   what legal problems will arise when the EPS is enforced for private and public building partners? Given the fact that the Urban Planning Act does not specifically endorse this instrument, some legal complications may be expected;

3   will the enforcement of the EPS appear as an additional bureaucratic burden? Application of the EPS requires substantial expertise, communication and control. It is questionable if the local government apparatus can handle this additional task. It is also possible that the bureaucratic burden overshadows the benefits of the instrument;

4   finally, attention must be drawn to the possible useful relation between the Environmental Performance System and the 'stolp method', which is discussed in another chapter. It is conceivable that the environmental aims of a certain 'stolp' of the city will have to correspond, or be compatible, with the EPL of a local land use plan in the same stolp.

It is obvious that Amsterdam still stands at the beginning of translating the idea of sustainable urban development into policies and applicable instruments. But an interesting move in urban planning has been made: a shift from environmental zoning as a means of relocating environmental burdens, towards a strategy for fostering sustainable urban development.

Note

1   Endre Timár is working at the Environmental Department of the City of Amsterdam, The Netherlands.

References

Ministerie van Verkeer en Waterstaat [Ministry for Traffic and the Maintenance of Dikes, Roads, Bridges and Navigability of Canals] (1990) *Tweede Structuurschema Verkeer en Vervoer* (SVV2) [Second Structurescheme Traffic and Transport], The Hague.
Physical Planning Department of Amsterdam (1994) *Globaal Bestemmingsplan IJ-oevers* [Broad Land Use Plan IJ-Banks], Amsterdam.
VNG [Association of Dutch Municipalities] (1992) *Bedrijven en Milieuzonering* [Industry and Environmental Zoning], VNG-publishers, The Hague.

Annex 10.1   Explanation of items in the EPL of the IJ Bank Project

The Floor-Space Index is a commonly used parameter to describe the density of building. A dense city gives bearing capacity for sophisticated environmental technologies, for excellent public transport, energy saving, multi-functional development and accessibility by foot/bicycle. Therefore, building in high densities is rewarded.

The (auto-)mobility problem is faced in a fivefold manner. A reasonable walking distance to public transport stimulates people to use it as well. Maintaining a small amount of parking places avoids triggering of the so called 'latent demand' of automobility and stimulates people to think of alternatives of the private car. On the other hand, incidental use of a car is almost unavoidable. Therefore alternative forms of car-ownership (car sharing), like 'call-a-car' systems are stimulated by giving physical space for their introduction. The use of the bicycle in Amsterdam is hampered by the high level of theft. Safe storing-places within walking distance will stimulate people to reconsider the use of the bicycle.

Noise nuisance eventually harms the image of 'living in a dynamic city', and people will try to move to the countryside. Reducing the noise level in a neighbourhood will add to the quality of city-dwelling and is therefore rewarded as well.

Sustainable (or 'ecological sound') building technology is relatively simple and reduces the environmental impact of urban development substantially. The municipality published an 'environmental preference list' in 1993. For each building element used nowadays, materials are categorised into 5 columns. Materials stated in column 1 are very environmentally friendly (and often very expensive); materials in column 5 are very harmful (and often inexpensive). Any use of these harmful materials is strongly discouraged in the EPS, and use of environmentally friendly materials are rewarded substantially.

New urban developments provide unusual opportunity for modern energy saving techniques like district heating and energy conserving architecture for offices and housing. Given the fact that district heating in a dense area is highly effective, the possible non-use of the technology is penalised.

Water saving is a comparatively easy measure. If the relatively clean rainwater from roofs is used for e.g. toilet flush, this is rewarded. Retention of water on the roof also diminishes the burden on the municipal sewage treatment plants. Application of water-saving equipment is rewarded as well.

One of the reasons that separated waste collection is not yet widely practised is the long distance people have to travel to recycling containers. Convenient location of waste collection stations is rewarded in this EPS.

Environmental features incorporated into this point scoring scheme are especially appropriate for developments in a dense urban setting, and will need to be adapted for other kinds of settings.

Annex 10.2   The Environmental Performance System

Environmental performance per planning zone in points		
	Performance	Points
THEME DENSE/COMPACT CITY		
Floor-Space Index in gross floorarea/lot:	> 2	1
	> 3	2
	> 4	3
THEME MOBILITY		
Walking distance in meters to nearest public-transport stop:	< 300	1
	< 200	2
	< 100	3
Parking standard for residential areas (no. of spaces/home):	≤ 1.0	1
	≤ 0.8	2
	≤ 0.6	3
Parking standard for work locations (no. of spaces per /250 m^2):	2 (B location)	1
	1 (A location)	2
	≤ 1	3
Parking spaces in built car parks, exchangeable for Call-a-car amenities (no. of cars):	> 5 cars	1
	> 10 cars	2
	> 20 cars	3
Secure bicycle storage within 100 meters (no. of places/home):	1	1
	2	2
	3	3

	Performance	Points
THEME NOISE NUISANCE		
Road traffic, noise level	max. exemption value	-3
	for each 3 dB(A) reduction	+1
	< preference value	+2
Rail traffic, noise level	max. exemption value	-3
	for every 3 dB(A) reduction	+1
	< preference value	+2
Industry, noise level	max. exemption value	-3
	for every 3 dB(A) reduction	+1
	< preference value	+2
THEME SUSTAINABLE BUILDING		
one or more elements realised with an alternative product from the BWA environmental preference list[a] (column):	column 5	-4
	column 4	-2
	column 3	0
50% of elements realised with an alternative product from the BWA environmental preference list[a] (column):	columns 1 and 2	+4
90% of elements realised with an alternative product from the BWA environmental preference list[a] (column):	columns 1 and 2	+7
THEME ENERGY		
Heating and/or cooling by means of cogeneration:	yes	3
	no	-1

	Performance	Points
Energy performance standard for utility buildings, according to NEN 2916 (in m^3 aeq/m^2 gfa):	< 30	1
Energy performance standard for homes (in m^3 aeq):	800	1
	750	2
	700	3
THEME WATER / GREEN SPACES		
Discharge of rainwater (from roof) to:	100% to sewer	-1
	<100% to sewer	+1
Compliance with recommendations on saving water[b]:	no	-2
	yes	+2
Planted roof or roof garden (not terrace):	yes	+1
THEME WASTE		
Access to waste-collection site (milieuabri) (max. distance from homes, in meters):	<100	1
MAXIMUM POSSIBLE SCORE:		47

a   For the BWA (soil/water/atmosphere) environmental preference lists, see Appendix D of the IJ riverbank land use plan (Physical Planning Department of Amsterdam 1994).

b   See parapraph 4.3 of the explanatory notes to the land-use plan (*Toelichting bestemmingsplan IJ-oevers*) (Physical Planning Department of Amsterdam 1994).

*Source:*Physical Planning Department of Amsterdam 1994

Chapter 11

A Method for Incorporating Environmental Aspects into Spatial Planning

N. Streefkerk[1]

Summary

'We can learn from the history of the Indians in North and South America, the small-holder on Java Indonesia, the public worker in The Bronx in New York, the manager who feels responsible for his firm and people, the spatial planner and the scientist who believe and work in terms of mutual respect.' This statement speaks to the mutual relationship or the interdependence of human beings and their natural environment. This ethical concept is not bound to a certain location and certain time, and applies to everybody all over the world.

The municipality of Zwolle, together with the Ministry of Housing and Spatial Planning and Environment of The Netherlands, has developed a method by which this concept can be translated into a practical instrument in which environmental values can be protected without overly constraining human land use. This instrument is called 'a Method for involving Environmental aspects into spatial Planning' or MEP. With this method, environmental influences caused by human activities, such as urban developments, are assessed in terms of the environmental sensitivities of the areas concerned, with the purpose of gaining insight into problems which might occur by implementing these developments. Practical experience has proven that the MEP works and can be applied very simply, not only within Zwolle but in other places within The Netherlands and in other parts of the world, such as the Third World. In The Netherlands the MEP has been published in a handbook or manual, which can be used by governmental employees dealing with the environment and spatial planning (Streefkerk 1992).

11.1 Introduction

Zwolle is a provincial capital with approximately 100,000 inhabitants, and is located in the east of The Netherlands. It is surrounded by areas of great natural beauty. The western part of The Netherlands is mainly urban and includes Amsterdam, Rotterdam, The Hague, and Utrecht (see Figure 11.1).

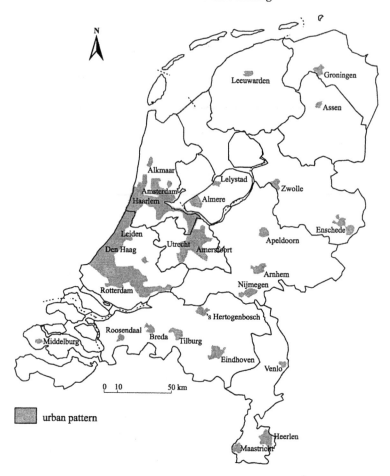

Figure 11.1   Urban patterns of The Netherlands

The policy of the Dutch government is to conserve the natural areas in the east of the country by discouraging urban developments within the small settlements there, and encouraging these developments in cities such as Zwolle, which will at the same time diminish urban pressure in the west of the country. In other words, Zwolle is considered an 'urban concentration centre', where the inflow of housing, tourism and industry will be encouraged. It is a fast growing city and presents the city council with a challenge: where to guide the location of these developments and where to protect the natural conditions in order to create sustainable living conditions for its inhabitants and protect its natural environment.

This sustainability is also a major item of the national and international policy and a major item of scientific research (Dutch Committee for Long-Term Environmental Policy 1994). Additionally, the National Environmental Policy Plan 'To Choose or to Lose' of The Netherlands in 1989 and Local Agenda 21 of the

UNCED conference in Rio de Janeiro in 1992 address this issue. Sustainability between man and his environment can be considered as the major aim of the MEP. In Zwolle, specific and practical problems posed several questions:

- Why should we submit an industry near a natural area to the same environmental regulations as a similar industry within the city itself?
- What is the best location for a new garbage treatment plant?
- What is the best route for the transport of hazardous products?
- How can we make an environmental map which can be used as an under layer for the making of a master plan for the city?

In order to cope with the political commitment to sustainable development, and with these specific questions, the City Council decided to develop a method for integrating environmental values into urban planning. With financial support from the Ministry of Housing, Spatial Planning and Environment the MEP was developed during 1990-1992, and presented in the form of a handbook (Streefkerk 1992).

11.2   Reasoning behind the MEP

The MEP can be considered as a reasoning process, which has been translated into a method, toolbox or instrument. This reasoning process includes the following points:

- spatial activities such as building houses, infrastructure and industry may cause negative environmental impacts such as pollution, noise, land removal, loss of nature, changing of ground water table, hazards, etc.;
- parts of the impacted environment have a varying degree of sensitivity to these influences. For example a housing area is sensitive to hazards with explosive products, while a grass field is less sensitive to explosions;
- comparison of the impacts which occur with these sensitivities reveals where there are risks of negative effects and where there are not, that is, where problems may occur and what kinds of problems these are. In the case of the housing area close to a chemical plant, for example, explosions can cause fatalities and damage to property, whereas in the case of the grass field near this plant this kind of risk is negligible;
- based on this information, the environmental manager suggests how these problems might be solved by zoning the urban developments and the sensitive environment, for example by creating zones for hazardous products and by zoning areas where people can work and live;
- finally this environmental proposal will be discussed with the planner or designer in order to make a definite plan or design which everybody agrees upon. In this design non-environmental aspects, such as economical values, are also taken into account.

This line of reasoning applies to existing situations, an existing situation which will be changed through redevelopment, or a totally new situation. In existing situations the existing impacts are compared with existing sensitivities. In the case of changing the existing situation into a new one, the new impacts and sensitivities are compared with the existing influences and sensitivities.

In the case of an entirely new situation, for example a proposed new housing development project with roads and industry, the new impacts from the roads and industries are compared with the new sensitivities of the residential development.

The MEP can be regarded as a toolbox for assessing combinations of existing and new impacts and sensitivities. MEP analyses can be done by means of sketches on paper or by mapping; sometimes computer assisted GIS is helpful.

11.3   Methodological Aspects

The MEP is based on regarding the environment as a system which consists of three main compartments (Tjallingii 1981; Odum 1983):

- the natural environment: soil, water, air, plants and animals;
- the non-natural environment: human artefacts and materials; and
- the anthropogenic environment: our society and the human members of it.

An area can be considered as natural when life under natural conditions is dominant, for example a primary forest, a natural lake, dunes, and deserts. Human beings are considered as natural as well. Non-natural is a situation where human artefacts are dominant and where plant and animal life including human beings are mainly absent, for example, a parking-lot, storage place of products and garbage. A human living area with houses and gardens is considered as a combination of natural and non-natural. Within this environment people play a 'double role', namely as active human beings who cause spatial, economical, social and environmental effects on their surroundings, and as receptive human beings, who have a sensitivity to these effects. In the MEP, the negative environmental effects such as pollution caused by human activities such as manufacturing and traffic are taken into account.

Figure 11.2 is an illustration of the definition of 'environment' and the role of people within this environment. The frame around the figure symbolises the area in which the environmental problems take place.

Sources of Environmental Impact

Human activities are the major source of environmental effects, and these sources include: manufacturing plants, agricultural land uses, logging and clear cutting of forests, infrastructure such as highways and power lines, power plants, airports and airstrips, housing, recreation and tourism, and specific activities such as landfill and quarrying.

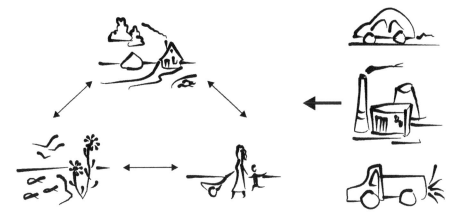

Figure 11.2   The concept environment

Environmental Effects

The MEP considers the following environmental impacts: local air pollution, smell, dust, soil and water pollution, noise with 50 decibel and 65 decibel, risk of explosions, risk of chemical calamities, removal or degrading of land and natural or non-natural vegetation, and impacts on ground and surface water. Environmental zoning considers only local effects, defined by the range or distance from their sources, which can vary from zero to approximately 300 to 1,000 meters (1,000 to 3,000 feet).

In The Netherlands, industrial sources have been classified based on their performance characteristics and the spatial extent of their impacts (VNG 1992). For example, the category 'bakery causes smell', while an 'oil refinery presents risk of explosions and causes air pollution'. General knowledge about other sources (RIVM 1991) makes it possible to give every kind of source a characteristic impact pattern. For example a highway with more than 25,000 traffic movements per day causes noise of more than 50 decibel over a distance of approximately 150 meters and causes local air pollution over a distance of approximately 50 meters.

In strategic planning (Hickling 1979; Archibugi 1994), there are two ways of approaching negative environmental effects: a) problem oriented, what are problems we might have to deal with and b) target oriented, how do we solve the problems? In the first case it is possible to have a broad view of all the negative effects we are dealing with and which might cause problems. In that phase of the planning process, exact information for each effect is not necessary and rough estimates are sufficient. Practical experience has proven that many environmental problems already can be solved in that phase of the planning process, without much difficulty. With the help of general knowledge of the sources and the estimated effects, it is possible to draw or sketch these influences on a map.

In the second case, it is important to know which sources and negative effects are causing the real problem, and which problems cannot be solved without great effort. This requires specific information about the negative environmental effects, which must then be calculated or measured. Financial and technical steps (commitments) have to be taken in order to reduce these impacts so that environmental standards can be met.

The MEP has been designed as a problem oriented method to reduce the difficulty of applying it. Figure 11.3 is an illustration of how estimated environmental impacts within an urban area can be drawn on a map and which of these impacts should be assessed by means of calculation or measuring.

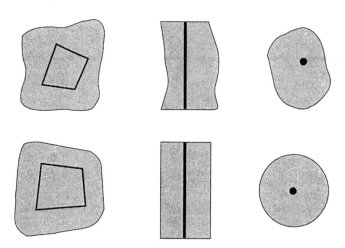

Figure 11.3   How influences can be mapped

Sensitivities of Environmentally Impacted Areas

The degree of sensitivity of the impacted area depends on the degree of naturalness of that area (Streefkerk 1992). Generally speaking, areas with a high degree of naturalness are highly sensitive, and areas with a low degree of naturalness are not sensitive. In other words, in areas where soil, water, and plant and animal life are dominant, the degree of sensitivity is high. In areas where artefacts are dominant this sensitivity is low. This is especially relevant to influences such as pollution, degrading of land and changing water regimes.

This is due to the fact that human activities in most cases lead to the introduction of artefacts and decrease of natural abiotic conditions, assuming that activities based on restoration and introduction of nature and natural conditions do not take place, as discussed later. The presence of human beings as biological creatures plays a special role in estimating the degree of sensitivity. Because of recognised regulations, in areas where the presence of people is dominant, the degree of sensitivity for noise and hazards is high, because people are considered to be sensitive to these influences. However situations with original wildlife are also considered as very sensitive to noise caused by people.

Using the MEP a classification of areas with the different functions and forms of land use has been made based on the degree of naturalness and characteristic degree of sensitivity of these functions to various environmental impacts. Figure 11.4 is an illustration of how these functions and their sensitivity can be drawn on a map. It is relatively simple to make a map of relative sensitivities, because it is easy to distinguish different functions in terms of pure abiotic with natural life, nature with artefacts, pure artefact, with human beings, without human beings etc. In Zwolle conventional city maps, topographical maps and aerial photographs are used for making the environmental map with sensitivities.

Comparison of Negative Environmental Effects and Sensitivities

Comparison of the map of environmental impacts with the map of sensitivities provides the basis for assessing the existing or new environmental situation of the area in terms of a range from 'good' or 'without problems expected' to 'poor' or 'problems expected'.

It is also possible to use ratings of the acuteness of the problems, from 0 to 9. Maps shaded from light coloured to dark coloured can be made. Generally speaking, situations with no impacts are related as 'good', situations with negative impacts but with a low degree of sensitivity for these impacts, such as areas with low naturalness, are rated 'fair', and situations with impact and with a high degree of sensitivity for these impacts, such as those in natural condition, are rated as 'poor'.

To illustrate: noise on a parking lot – 'fair'; noise on a housing or wilderness area – 'problems expected'; air pollution on a parking lot – 'fair'; pollution of a nature area – 'problems expected or poor'. Pollution in the form of soil or water contamination is always considered a poor situation, however a gradation can also be made, for example used oil spills on bare land is worse than used oil on a paved area. Combinations of the number of kinds of impact and degree of related sensitivities we are dealing with can be expressed in a range of resulting conditions from 'no problems expected or good' to 'fair' to 'problems expected' or 'poor', or by means of rating values from 0 till 9.

Table 11.1 shows a table of such values concerning resulting conditions. On the vertical axis, at the left side of the table, the number of impacts is shown; on the horizontal axis, across the top, the degree of sensitivity ranges from very sensitive, sensitive, to less sensitive, and on the vertical axis on the right side the rating for resulting condition and in terms of 'good' to 'poor' is shown. With no impacts, the table shows 0, or good. With the combination of one impact on a less sensitive area, the rating is 1. With the combination of one impact on a sensitive area, the rating is 2. With the combination of one impact on a very sensitive area, the rating is 3. With the combination of two impacts both on a less sensitive area, the rating is 2. The combination of two inputs with one on a less sensitive area and one on a sensitive area, the rating is 3, etc. In a situation with three impacts on an area which is sensitive to all of these impacts, the table shows the rating of 9.

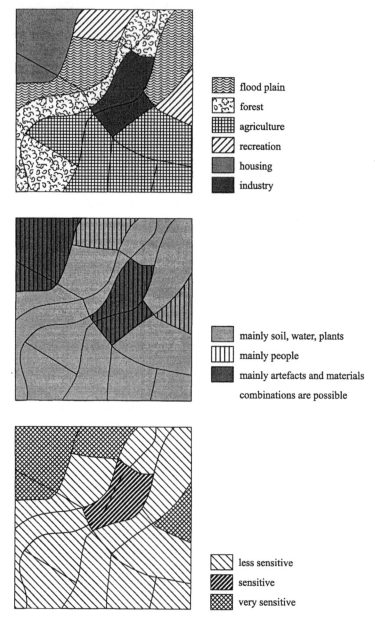

Figure 11.4   Spatial functions and their degree of sensitivity for explosions

Environmental View

An integrated approach to assessing environmental impacts by means of a MEP is aimed at offering options for solving these problems. The overall goal is to determine how much can be accomplished for the environment, without blocking or frustrating spatial or urban developments unnecessarily. The product of this analysis looks like a design or sketch, if necessary with points noted for further investigation. This design or sketch includes various proposals for avoiding undesirable environmental conditions, including:

- locations or zones proposed for spatial or urban developments;
- zones where environmental values have to be protected, and
- different measures which can be taken to mitigate negative effects.

Table 11.1 Confrontation between number of influences and the degree of sensitivity for their influences

Degree of sensitivity → / Number of influences ↓	Less sensitive	Sensitive	Very sensitive	Final Score	
no influences	0	0	0	0	good
1 influence	1			1	
		2		2	
			3	3	
2 influences	1.1			2	
	1	2		3	
	1		3	4	
		2.2		4	
		2	3	5	
			3.3	6	
3 influences	1.1.1			3	
	1.1	2		4	
	1.1		3	5	
	1	2.2		5	
	1	2	3	6	
		2.2.2		6	
	1		3.3	7	
		2.2	3	7	
		2	3.3	8	
			3.3.3	9	poor

Examples of these measures are: the creating of buffer zones between sources of pollution and sensitive areas, shielding such as tree belts or walls to lessen the noise on houses or to protect natural areas, sound insulation of buildings, creating natural corridors between natural areas, and making pavements or liners to prevent seepage.

Integrated design has used several of these measures for about three decades (McHarg 1969). After the MEP analysis and the proposals have been publically discussed, these points that are agreed to can be included in the local general plan.

11.4   Synopsis of Projects

As mentioned earlier, the MEP can be used for a wide range of spatial problems. In scale, these can range from 1:1,000 to 1:100,000 (from approximately 1 inch to 100 feet to 1 inch to 10,000 feet). It can also be used for both urban and rural planning, and from master planning to site planning. Projects within Zwolle which illustrate this range of uses include:

- the environmental zoning of an urban fringe project with housing, offices and industry near shunting-yards and a heavy metallurgical industry, 'where can we find the most suitable zones';
- an Impact Assessment for the location of 10,000 houses in a rural area, 'which locations are suitable and what is the most proper outline of this housing development plan' (see the example below);
- a zoning plan for an industrial area, 'where to locate chemical, transport and smaller industries'; and
- a general environmental map for the municipality of Zwolle.

Elsewhere within The Netherlands, the MEP has been used for projects such as:

- environmental mapping for urban and industrial development of the city of Groningen;
- a land re-allotment plan (nature, agriculture, tourism and recreation) for the rural area around the city of Arnhem; and
- an environmental checklist (matrix) of environmental impacts and sensitivities used by the city of Amsterdam.

Two projects in the Third World using the MEP methodology are:

- a master plan for land use in valleys in the Andes in Bolivia, 'how sensitive are slopes for grazing and settlements';
- a land development plan for settlements and land use in the Dominican Republic, 'how sensitive are areas to irrigation measures and a new infrastructure' (Streefkerk 1986).

11.5   Description of an Example

As noted earlier, Zwolle is a rather small but fast growing city. One of the projects the city was dealing with was finding and designing a site for 10,000 houses, with the least amount of harm to the environment. The impact assessment and planning process consisted of three phases:

1   an inventory of the whole municipal area to determine zones in which, technically speaking, 10,000 houses can be built;
2   compare these zones on an broad basis in order to find out which zone or combination of zones is the most feasible for the houses;
3   make the site design for the zone which is chosen.

Phase 1 of the planning process was problem oriented and resulted in identifying three zones for further investigation (see Figure 11.5). During this phase, with help of the MEP, an integrated environmental zoning map was used as a tool on which to base this decision.

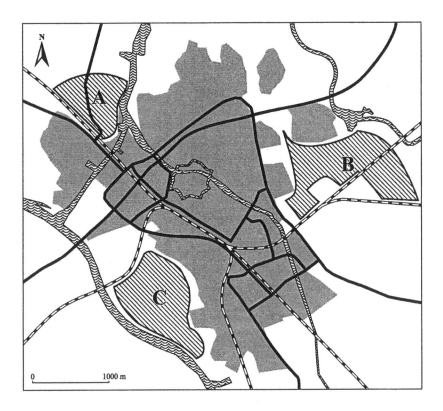

Figure 11.5   The city of Zwolle with three alternative locations A, B and C

Phase 2 was also problem oriented, because the three zones provided enough space for locating the housing development area. However during this phase there were uncertainties on particular areas and more detailed research was necessary, to be sure that no constraints would emerge later on. It was possible to choose one of the three locations. The MEP and some extra research on soil contamination and noise were used to include environmental concerns into the planning process and to facilitate making a choice.

Phase 3 was target oriented because it had to result in a final design and local allocation plan for building the new houses. In this phase, some environmental aspects have been studied in detail, to be sure that legal standards will be met. Important aspects were noise, soil contamination, existing natural vegetation, and additional aspects such as saving energy, the ground water situation, and efficient means of accommodating traffic flows, including bikes, public transportation and garbage collection. In phase a MEP analysis was not necessary.

In phase 2, the MEP analysis was used to identify and map the three locations, as illustrated in Figure 11.5. Existing sensitivities, and new sensitivities and negative impacts were analysed for each location. Areas with negative effects were: not suitable for housing, because existing sensitivities which will be destroyed or disappear as the result of the new development project itself; not suitable, because existing sensitivities in the area surrounding the site will be affected by the new development project; or not suitable because existing environmental impacts will effect the new sensitivities of the housing area. Areas where these effects will not occur are residuals of the process and can be considered as suitable.

Additionally, each location was analysed to determine which offered the best opportunity for mitigation of negative effects, and where extra research is necessary. Of the three options, that location which was found to have the least negative effects or largest suitable area, expressed in square meters. Table 11.1 shows a table of such values concerning resulting conditions. On the vertical axis, at the left side of the table, the number of impacts is shown; on the horizontal axis, across the top, the degree of sensitivity ranges (square yards), and the best opportunity for mitigation, is the most favourable from an environmental point of view. These environmental assessments were combined with other criteria, for example economic implications, after which the choice among the three locations was made by the City Council.

Figures 11.6 and 11.7 show how the MEP analysis and environmental assessment for one of the locations, in this case option A, was done. This environmental approach has also been used as a basis for the final plan or design in Phase 3 of the project.

11.6   Conclusion

The MEP has been developed from professional practice experiences such as issuing licences for factories and developing urban plans. It is an attempt to link a practical bottom-up approach and scientific methodology.

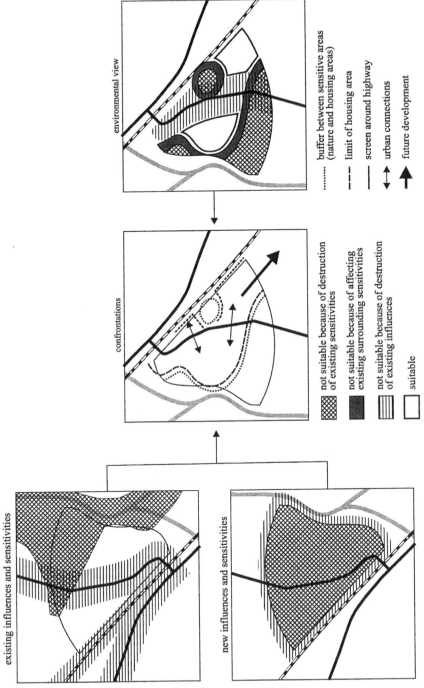

Figure 11.6 Analysis of option A from Figure 11.5

The MEP has been written especially for the situation in which we do not need to know all the detailed information before appropriately taking action. Our environment cannot wait for that. Much can be done without having available all the detailed information. However, the MEP is an instrument to identify and assess environmental problems and to identify detailed issues, which need further research.

Environmental problems in the Third World can have greater impacts on human and natural life than in the Western World, even in situations where there is no political demand to solve these problems. The MEP is an instrument which can be applied in the Third World, especially since it may not require detailed information. The ethical concept of treating our fellow human beings and our environment with respect, as mentioned at the beginning of this chapter, implies an integrated approach to spatial and environmental problems. Integrated, in this context, means considering sensitivities and intrusive activities without prior judgement. Employing this approach can facilitate constructive negotiations, leading to widely accepted solutions.

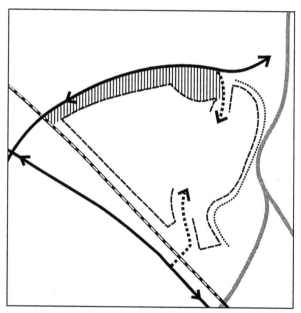

Figure 11.7   Final Plan (see Figure 11.6 for description symbols)

Note

1   Nico Streefkerk is working at the Department of Environmental Affairs, Municipality of Zwolle, The Netherlands.

References

Archibugi, F. (1994) The basic issues of ecological city planning, International Symposium on Urban Planning and the Environment, Seattle, WA, 2-5 March, 1994.

Dutch Committee for Long-Term Environmental Policy (1994) *The Environment: Towards a sustainable policy*, Kluwer Academic Publishers, Dordrecht, Boston, London.

Hickling, A. (1979) *Aids to Strategic Choice*, Tavistock Centre, Belsize Lane, London.

McHarg, I. (1969) *Design with Nature*, Natural History Press, Doubleday and Company Inc., Garden City, NY.

Odum, H.T. (1983) *System Ecology*, Wiley, New York, NY.

RIVM [The Dutch National Institute for Public Health and Environment](1991) *Nationale Milieuverkenning 1990-2010* [National Exploration of the Environment, chapter 8], Samson H.D. Tjeenk Willink bv., Alphen aan de Rijn, The Netherlands.

Streefkerk, N. (1986) Landinrichting en ecologie in de Derde Wereld [Physical planning and ecology in the Third World], *Landbouwkundig Tijdschrift* [Agricultural Magazine], Vol. 98, No 2.

Streefkerk, N. (1992) *Handboek Beoordelingsmethode Milieu* [handbook for using the MEP], VNG-Publishers, The Hague.

Tjallingii, S.P., and A.A. de Veer (eds) (1981) *Perspectives in Landscape Ecology*, Pudoc, Wageningen, The Netherlands.

VNG [Association of Dutch Municipalities] (1992) *Bedrijven en milieuzonering* [Environmental Zoning and industries], VNG-Publishers, The Hague.

Chapter 12

Innovative Mitigation at a Treatment Plant: The Applied Waste Water Technologies Research Program

G.M. Bush[1]

12.1 Introduction

In the minds of everyone, except perhaps a few sanitary engineers, waste water treatment plants are intrusive industrial facilities. They are also critical public works sites that are federally mandated and designed to protect public health and benefit environmental quality. The King County Department of Metropolitan Services (Metro) has been a leader in efforts to avoid and reduce conflicts between an environmentally intrusive facility, the West Point Treatment Plant, and environmentally sensitive land uses. In this example, the surrounding land uses are high quality single family housing and park land. The focus of this chapter is on impact mitigation efforts at this vital treatment plant, and on a new mitigation initiative Metro has undertaken: the Applied Waste Water Technologies Research Program.

12.2 West Point Expansion

The West Point wastewater treatment plant was constructed in 1966 as a primary treatment plant with an operating capacity to treat 125 million gallons per day (MGD). This treatment plant is located at the western edge of what was a major military reservation and now is Discovery Park. The decision was made in 1986 to expand treatment capability at the plant to a secondary level, as required by federal law. The plant discharges treated effluent to Puget Sound via a deep off-shore outfall. Expansion or the facilities at West Point, rather than construction of a new treatment plant at another site, was a hotly debated political issue that tended to galvanise opponents and supporters. Extensive environmental studies (Metro 1988) were completed to assess the potential impacts of the expansion at West Point and several other alternative sites. The final decision to expand the plant to secondary treatment was determined by a one vote majority on the 44 member Metro Council.

The debate generally focused on two issues; the higher cost of siting a new plant, and the potential environmental impacts at Discovery Park versus the disruption that would be created by construction of a new conveyance system to route waste water to a new plant. The potential for environmental impacts from 7 years of construction at West Point were widely acknowledged. Less certain was the significance of continued operation of the plant on aesthetics and enjoyment of Discovery Park.

Because waste water was discharged at West Point decades before the original plant was constructed, all major conveyance pipelines in Seattle are routed to the site. The cost of expansion at West Point is more than $500 million, yet this was the lowest cost alternative studied. Construction of a new plant at a suitable, industrially zoned site within the City would have added an estimated $350 million to the project's cost.

By looking at the site in an aerial photo you would see that West Point is a prominent point of land jutting into Puget Sound. The site is only 32 acres in size and consists primarily of fill that replaced a natural mud flat. The relatively small surface coverage or 'footprint' of the plant for its 125 MGD capacity is a key feature when considering the mitigation measures incorporated into the design.

The plant at West Point is bordered by Seattle's Discovery Park on the upland side and by the Puget Sound on the West and North edges. Discovery Park is the largest park in the City of Seattle and receives more than one million visitors each year. The park provides significant habitat for wildlife since it is largely forested with native vegetation. Discovery Park is home to many animal species, including a nesting pair of American bald eagles. The eagles have been monitored extensively, both prior to and during the construction period, to ascertain what potential impacts construction might have upon nesting, and help to design mitigation measures. The pair has successfully fledged offspring during each of the last 4 years of plant construction activity.

The marine shoreline at West Point provides rearing habitat for salmonid species in one of the largest eel grass beds in the central basin of Puget Sound. Underwater habitat disturbed during construction is being restored. The shoreline near the treatment plant is a popular destination for visitors to the park. Although the beach area is generally accessible only by foot traffic, and does not offer vehicle parking for the general public, it still rates highly in daily public use. Continued public access to the shoreline trail was a key objective during construction. Special features such as a pedestrian overpass have been provided to allow walkers to bypass construction activity.

12.3   Mitigation Aspects of the Project

Other mitigation features were incorporated in the plant design and construction process. Of the $500 million cost of the plant upgrade, nearly $90 million could be considered mitigation. However, that amount includes the cost of some design features that serve operational purposes as well as providing added impact mitigation. For example, the extensive retaining wall placed on the upland side of the plant enabled designers to minimise intrusion into Discovery Park, avoid fill on the shoreline side, and lower the profile of the plant from public view. A tall soil berm was constructed around the plant to further screen views from the park and beach area. The berm was completely landscaped with native species and incorporated unobtrusive security fencing to separate the treatment plant from public access areas.

As mentioned earlier, the plant is situated on only 32 acres, an extremely small site by U.S. standards. This presented some interesting engineering challenges. Operational features have been stacked in plan to minimise the footprint and the need for added pumping, since treatment facilities rely largely on gravity flow. In addition, Metro has made commitments not only to maintain the scale of the plant, but also to look at ways to reduce the footprint in the future. Air quality protection and prevention of odours was addressed with control technology. Extensive baseline air quality monitoring was performed to support Metro's application for air quality operating permits. Metro established an odour control policy that allows no more than three odour units to be detectable at the plant's fence line, which is close to the undetectable level. To achieve this standard, control measures such as carbon scrubbers have been placed at odour producing locations in the plant. Many features of the treatment plant process are also covered to control air flow from the site.

Other mitigation commitments included restrictions to the amount of truck traffic that will come and go from the facility. This plant currently processes waste water solids by anaerobic digestion. Digesters are large tanks, and are one of the more prominent features of this and many treatment plants. The sludge treated in these digesters, which we refer to as biosolids, is hauled off site in large dual trailer trucks and land applied as a soil conditioner. Metro has placed limits on the number of truck trips that can enter and leave the plant each day to minimise impacts to park users.

In addition to the measures presented, Metro has provided compensatory mitigation to further offset the loss of shoreline land for other uses and general disruption to the surrounding community. The Seattle Shoreline Improvement Fund was established with a $25 million contribution. The fund will be used to acquire and enhance other shoreline property for public benefit.

Despite the wide range of mitigation planned for the project, construction was almost delayed by a late citizen challenge to the construction permits. Four community based interest groups brought suit to delay construction citing a variety of environmental concerns and seeking added mitigation through the City permit process. Metro, faced with the prospect of imposing a moratorium on sewer hook-ups in the service area, and being fined by the state for missing schedule milestones, negotiated an out of court settlement that allowed the project to begin construction on the intended schedule. This settlement agreement (Metro 1991) was the basis for formation of the Applied Waste Water Technologies Research Program (AWT).

12.4 The Applied Waste Water Technologies Research Program

As part of this unique settlement agreement, Metro agreed to budget $5 million to research the application of new technologies that offer opportunities to reduce environmental impacts at the West Point plant. In making this commitment, the agency recognised that there may be broad regional or agency wide benefits to the waste water system. An interdisciplinary task force that includes representatives from engineering, process control, community relations, and other disciplines was formed to staff the effort. The task force reports to a management steering committee and receives guidance from a Citizens Advisory Committee. The advisory committee includes representatives from each of the interest groups that was party to the West Point settlement agreement, as well as others interested in waste water technology. The AWT program has been in place since early 1992. Although the program was born out of a conflict, it presents an exciting opportunity to explore innovative new research projects.

Because the AWT is focusing on technology that can be applied today to reduce key environmental impacts it will not undertake traditional research and development activities. We do not have either time or budget to spend 20 years or more developing concepts from the laboratory scale. Instead the focus is on emerging technologies that can perhaps benefit from Metro's monetary support and engineering expertise. The AWT is placing a serious emphasis on reducing the key impacts that are of most concern to surrounding residential communities. These key impacts were addressed briefly in the settlement agreement and include the area required for solids handling facilities (the footprint), odour control measures, and reducing truck traffic. The AWT will also pursue activities aimed at reducing other environmental impacts, including promoting energy conservation, reducing noise and improving aesthetics.

To help focus the research program's objectives staff initiated a series of technology and impact screening workshops. Citizens were invited to participate. We wanted to create a broader understanding that trade-offs exist between solutions to environmental impacts. The program had some success in asking participants to prioritise environmental impacts. From these workshops we gained a better understanding that citizen concern centred on the presence of digesters at the plant, and the perception of odours associated with both the solids handling process at the plant and the use of trucks for hauling treated biosolids. Also, the potential cost of new treatment technologies was a key concern expressed during the workshops, as was the need to maintain all existing mitigation commitments. There continues to be some public interest in the ultimate removal of digesters from the treatment plant. If a suitable technology for solids processing can be implemented, the land occupied by the digesters could possibly be returned to public access. The settlement agreement included an added monetary incentive to encourage Metro to continue to work to towards this goal. If digesters can not be removed by 2005, Metro will contribute an additional $1 million to the Shoreline Improvement Fund for each acre occupied by the digesters.

With the help of consulting engineers from Montgomery Watson, a detailed assessment of solids processing technologies was prepared (Montgomery 1993). Through the workshop process, these technologies were screened to identify techniques that would meet our key impact reduction objectives. The AWT is undertaking research activities to further refine our knowledge of solids processing. First, the AWT has begun a phase of pilot testing of specific solids processing technologies. In addition, we are continuing research on other proposals with academic institutions, private consultants and Metro staff. Until 1995 we have initiated about six research projects and are currently negotiating contracts to conduct pilot tests of solids processing technologies at Metro's East Division Reclamation plant. In 1995 our goal was to stage five pilot test projects over the following three years.

The initial pilot test projects will look at several innovative processes. One test will attempt to enhance the solids treatment process using a technology known as anoxic gas flotation (Burke 1993), which is a modification of the conventional anaerobic digestion process. If successful, this test may demonstrate that digesters can be made to operate more efficiently, ultimately requiring less space for tanks, producing more reusable methane gas, and reducing odour emissions. Our second pilot test will construct a digester in a 500 foot underground shaft. The deep shaft digestion process (Alpert and Cuthbert 1993) offers promise for significantly reducing the footprint of facilities on the surface, while producing a heat treated biosolids product that can still be used in land application. Temperatures created in this digester will exceed those typically found in conventional anaerobic digestion, providing the added benefit of producing a product with reduced pathogens. Some of the other AWT research activities include the following:

- the Biofilter Development Project;
- a study of struvite formation in biosolids;
- high efficiency dewatering assessment;
- West Point biosolids truck odour evaluation;
- dehalogenation of organic pollutants in anaerobic digestion.

Descriptions of each of these projects are found in the semi annual report prepared by the AWT (Hummel 1993). Each of these projects has been designed to meet the overall program objective of impact mitigation.

Upcoming AWT activities will move into new areas. If the pilot tests are successful the AWT may conduct larger scale demonstrations of the technologies. One new solid processing technique (the so-called VerTech Treatment System) that we are hopeful to pilot test will become operational at a new plant in The Netherlands. The technique reduces sludge to an inert ash product via wet oxidation. This technique of wet oxidation was tested at a facility in the United States more than 10 years ago, however no operational scale treatment plants have incorporated this process until the new plant in the Netherlands. This technology also uses a deep underground shaft to house most of the process facilities. The typical shaft may be as deep as 5,000 feet and would be constructed using oil well drilling techniques. Solids are added at the top of vessels placed in the shaft and are pressurised as they fall. Oxygen is added and the sludge is oxidised producing an ash slurry. Substantial quantities of heat are released by the process and can be recaptured and put to beneficial use. In addition, the ash product has applications as an additive in construction materials. Since large volumes of air are not released in the process, air pollution control requirements are minimal.

Another new initiative that we are excited about is a collaborative program with the University of Washington College of Engineering that we are calling the Metro Fellowship Program. Under this research and training scholarship program Metro will fund research scholarships for up to 3 graduate students per year during a 3 year trial period. Students will apply for scholarships and will be selected for the program by University faculty in conjunction with Metro approval. They will target their thesis research to meet the major objectives of the AWT, i.e., reducing the environmental impacts of physical facilities and operations at the West Point treatment plant. The fellowship program is designed provide research and training experience for graduate students in environmental engineering. In addition, it provides an opportunity for Metro and the University to collaborate on research and development of new technologies for implementation at Metro facilities.

We are very encouraged about this program and hope that if successful it could be expanded to include other disciplines, and perhaps other universities, with co-ordinated funding coming from a variety of sources. The approach of targeted research has been applied successfully by private industry, but this fellowship program may be one of the first of its kind initiated by a government agency like Metro.

Over the course of the next 5 years, the AWT hopes to explore a variety of new topics. We are making progress toward an agreement to host a major demonstration of fuel cells, used to produce electricity, with a local utility and private manufacturer. This demonstration project would use digester methane gas produced at Metro's East Division Reclamation Plant, as a hydrogen rich fuel source in a molten carbonate fuel cell. The cell would produce one megawatt of power. We believe fuel cells hold considerable promise for high efficiency power generation. They have several unique advantages over conventional power production techniques, e.g., internal combustion engines. Fuel cells are small in scale and produce very low emissions. They also allow new power sources to be placed where they are needed, rather than relying on transmission lines. The AWT team has visited operational fuel cells in the Los Angeles area. Even in this region known for its extensive regulation of air pollution sources, new fuel cell power plants can be sited without the typical air quality regulatory permits.

Another topic of interest to the AWT is the possible the use of alternative fuel vehicles at our plants to reduce air and noise pollution. In addition, we hope to study the effects that garbage grinders ultimately have on treatment plant sizing, by directing food wastes to the wastewater treatment plant instead of to recycling and other disposal options. Prospects for the reuse of treated effluent is another area that holds promise of further impact reduction, not only in protection of receiving water quality, but also reducing industry's reliance on potable water sources. Metro has several demonstration projects underway that use reclaimed water for supply and district heating.[2]

12.5   Conclusions

The West Point waste water treatment plant is a prime example of a large industrial facility located next to an exceptionally environmentally sensitive land use, in this case a large park. The West Point case demonstrates a number of innovative ways to limit the effects of various processes employed to the boundaries of the site; that is, to internalise externalities. This case also represents an expanded view of how to mitigate spillovers, including an applied research program aimed at further improving performance, and off-site mitigation in the form of funding shoreline improvements in other areas. These features of the West Point project should provide ideas useful in informing efforts to deal innovatively with the issues of spillovers from other sorts of large public and private sector projects in urban contexts.

Notes

1   Gregory M. Bush is working for the Planning and Real Estate Division, Metro, in Seattle.
2   Water re-use demonstrations conducted by Metro have focused on three uses of highly treated effluent: landscaping irrigation; street and sewer cleaning; and district heating/cooling (which seeks to use the residual heat value of the effluent for heating and cooling applications in industrial and manufacturing facilities).

References

Alpert, J., and L. Cuthbert (1993) *Proposal to Metro*, E&A Environmental Consultants Inc. and Deep Shaft Technologies Inc., Bothell, WA.
Burke, D. (1993) *Proposal to Metro*, Western Environmental Engineers, Olympia, WA.
Hummel, S. (1993) *AWT Program Activities Report*, Seattle, WA.
Metro (1988) *West Point Secondary Treatment Facilities, Final Environmental Impact Statement*, Seattle, WA.
Metro (1991) West Point Settlement Agreement, Seattle, WA.
Montgomery, J.M. (1993) *Wastewater Solids Treatment and Processing Technology Assessment*, Consulting Engineers Inc., Bellevue, WA.

Part C
Integrated Environmental Zoning as a Sophisticated Instrument to Deal with Environmental Spillovers

Chapter 13

Dutch Integrated Environmental Zoning: A Comprehensive Program for Dealing with Urban Environmental Spillovers

D. Miller[1]

13.1 Introduction

As with most western countries, the Dutch national government has adopted a number of environmental programs, beginning with the Pollution of Surface Waters Act and Air Pollution Act in the 1970s, and followed by the Soil Pollution Act. However it was soon realised that the requirements of these programs overlapped in part, and that sometimes means to resolve one kind of environmental problem created other forms of pollution. Additionally, since different governmental agencies had responsibility for administering these various regulations, industry found it confusing and time consuming to obtain all of the necessary permits to function or to expand production. In fact, objections by industry were a major stimulus to the government to streamline and simplify the permit application and processing procedures.

Consequently, the Environmental Protection Act of 1980 sought to co-ordinate this process, in part through an improved procedure for environmental impact assessments. A second and major step was taken with adoption of the national Environmental Management Act, which explicitly recognised the need for dealing in an integrated manner with environmental protection, rather than regulating various threats separately (VROM 1991a, p. 6).

A provision of the Environmental Management Act mandated the development of the National Environmental Policy Plan (NMP), also known as the Green Plan, which has received international attention because its major purpose is to control environmental problems in The Netherlands within twenty-five years (VROM 1992, p. 19). An important provision of the NMP is a commitment to design and test a new approach to dealing with environmental problems in urban areas: Integrated Environmental Zoning IEZ). Integrated Environmental Zoning is an innovative effort to account for several environmental spillovers from manufacturing activities, and to manage their impacts on surrounding residential areas. The five specific environmental impacts which IEZ addresses are noise, odour, carcinogenic and toxic forms of air pollution, and external safety in the form of fire and explosion (VROM 1991b, p. 150). This program called for experimental application of the provisional system for this new program in a number of pilot projects, located throughout The Netherlands, to test the design of this approach and to determine needed changes before it is enacted as a requirement of all local governments within the country. These pilot projects are now underway. Several chapters in this volume report on findings and conclusions from some of these pilot projects. Two other chapters present alternative approaches being tried in Amsterdam, which may provide useful modifications to the provisional system for IEZ. The major purposes of this chapter are to describe the provisional system for IEZ in operational detail, to critically assess this methodology, and to provide an overview of some of the experience generated by the pilot projects.

13.2   The Provisional System for Integrated Environmental Zoning

In 1990, the Dutch Ministry of Housing, Physical Planning and Environment (VROM) published the Ministerial Manual for a Provisional System of Integrated Environmental Zoning (VROM 1990). Major features include a set of quality standards for the five forms of environmental spillover which are the initial focus of the program, a measurement scale for these, and a technique for calculating a cumulative index value for the impact that these five forms of pollution have on sites adjoining or near manufacturing activity (Table 13.1). The provisional system also includes a set of pollution abatement measures and land use regulations associated with the various categories of these cumulative pollution scores, which are intended to reduce or eliminate unacceptable environmental spillovers on residential areas.

The provisional system differentiates between source oriented and effect oriented actions to reduce the impacts of pollution. Source oriented measures seek to reduce emissions produced by manufacturing establishments, which are referred to as environmentally intrusive activities, and are the preferred means of resolving environmental problems. Effect oriented measures are intended to protect residences, referred to as environmentally sensitive functions, from impacts through planning their location at safe distances from manufacturing activities, and even relocating households from areas which can not be cleaned up enough to meet the standards specified by the provisional system.

Environmental Categories and Classes

Environmental quality standards are set for each of the five categories or kinds of pollution and hazard. These standards are derived from prior environmental policy and, in the case of noise, from previously enacted legislation. Each of these standards includes a limit which is a level of impact considered threatening to health or of unacceptable nuisance to residents, and a target which designates an acceptable level of environmental quality for residential areas. Tables 13.2 and 13.3 show that each of the five environmental categories (kinds of pollution) are divided into five classes of environmental quality bounded at the high end by the unacceptable level or limit (Class E), and at the low end by the desirable or target level (Class A). This provides a scale of measurable environmental impact (or load) for each category of pollution.

Table 13.1 Sectors, categories, components and agents chosen for the provisional system

Sector	Category	Agent component
Noise	Industrial Noise	Impulse noise Tonal noise Low-frequency noise
Air Pollution	Odour	-
	Toxic substances	Tetrachloroetene Trichlorothene Trichloromethane (=Chloroform) Phenol Styrene Tetrachloromenthane Dichloromethane Toluene
	Carcinogenic substances	Ethylene oxide 1,2-Dichloroethane Acrylonitrile Propylene oxide Epichlorohydrine Benzene
External Safety	Major industrial hazard	Risks from explosive substances Risks from (in)flammable substances Risks from toxic substances

Source: VROM 1990

These two tables also indicate that the provisional system treats existing and new situations differently, and that the standards for new development are more demanding. Existing situations include manufacturing firms operating before the provisional system was established, and housing which was present or under construction before the new regulations, as well as replacement housing (VROM 1990, Appendix 13.2). New situations refer to new plants or expansion of existing firms which may intensify impacts on surrounding areas, as well as new housing proposed for location in impacted areas. Standards for new development are provided in Table 13.3, which indicates that residential development is more restricted and that abatement of existing pollution is more demanding than for existing situations (Table 13.2). While the impact classes used by these two tables are the same, the upper limit for existing situations (Class E, Table 13.2) is higher than for new development (Class D, Table 13.3). Areas with Class A environmental conditions are desirable for new construction of homes, but the intermediate Classes B through C represent pollution levels which while not desirable for residences may be accepted temporarily under certain circumstances.

Since the spillovers addressed by IEZ tend to diminish with increasing distance from the source, mapped zones based on environmental impact improve for areas increasingly remote from intrusive activities. These distances can be reduced when pollution is reduced at the source. Physical planning plays an important role in the IEZ program by designating only those areas at safe distances from sources of pollution or hazard for residential development.

Table 13.2   Existing locations

Category	Class E	Class D	Class C	Class B	(Class A)
Industrial noise in dB(A)	> 65	65 - 60	60 - 55	< 55	(< 50)
Major industrial hazard in individual mortality risk per year	> 10^{-5}	10^{-5}-10^{-6}	10^{-6}-10^{-7}	< 10^{-7}	(< 10^{-8})
Odour in odour units per m^3 as a 98-percentile	> 10	10 - 3	3 - 1	< 1	(< 1^*)
Carcinogenic substances in individual mortality risk per year[**]	> 10^{-5}	10^{-5} - 10^{-6}	10^{-6} - 10^{-7}	< 10^{-7}	(<10^{-8})
Toxic substances as percentage of the No Observed Adverse Effect Level	> 100	100 - 10	10 - 3	< 3	(< 1)

[*]   As a 99.5 percentile
[**]   A maximum individual risk of $x.10^{-6}$ applies cumulatively to an x-number of carcinogenic substances, with a maximum of 10^{-5}

Source: VROM 1990
Note: Classification of sectoral environmental loads based on (draft) standards and supplementary chosen values. Class A in parentheses has no significance in existing situations.

Summarising Environmental Impacts

Appraising the overall environmental impact from the environmental categories on portions of the surrounding area involves combining the measures for these five kinds of spillover into a summary figure. Since there is no scientific basis for weighing these categories in terms of their undesirability, the decision was made to use a system of standards to '...assign consequences to the presence of several non-negligible loads' (VROM 1990, p. 14). This was facilitated by defining the scale for each of the five categories in a comparable manner: as ranging from the acceptable level (Class A) to the unacceptable level (Class E). Tables 13.2 and 13.3 present the measures of impact specifying each of these five classes, for each of the five kinds of pollution. As Tables 13.4 and 13.5 indicate, various combinations of Classes A through E are used to arrive at six integrated ('integral') classes, numbered I through VI, which specify the level of total environmental load at each location within the area surrounding a source of pollution. Class I locations are mapped as white areas, representing portions of the study area in which environmental quality is acceptable for residential use. Class VI locations are mapped as black areas, indicating that the combined impacts are greater than the maximum allowed for residential use. The intermediate integrated classes II through V are mapped in increasing values of grey to represent locations in which the total environmental load is undesirable, but permitting some land use flexibility. Note that the specifications for determining the integrated classes presented by Table 13.4 deal with existing situations as defined earlier, and employs the subscript B for each of these classes, differentiating these from the counterpart integrated classes for new situations (Table 13.5) which are more restrictive and which are designated by the subscript N.

Table 13.3   New situations

Category	(Class E)	Class D	Class C	Class B	Class A
Industrial noise in dB(A)	(> 65)	> 60	60 - 55	55 - 50	< 50
Major industrial hazard in individual mortality risk per year	(> 10^{-5})	> 10^{-6}	10^{-6}-10^{-7}	10^{-7} - 10^{-8}	< 10^{-8}
Odour in odour units per m^3 as a 98-percentile	(> 10)	> 3	3 - 1	< 1	< 1[*]
Carcinogenic substances in individual mortality risk per year[**]	(> 10^{-5})	> 10^{-6}	10^{-6} - 10^{-7}	10^{-7} - 10^{-8}	< 10^{-8}
Toxic substances as percentage of the No Observed Adverse Effect Level	(> 100)	> 10	10 - 3	3 - 1	< 1

[*]   As a 99.5 percentile
[**]   A maximum individual risk of x.10^{-6} applies cumulatively to an x-number of carcinogenic substances, with a maximum of 10^{-5}

Source: VROM 1990
Note: Classification of sectoral environmental loads based on (draft) standards and supplementary chosen values. Class E in parentheses has no significance in new situations.

Mapping Integrated Classes of Impact

Figures 13.1 and 13.2 illustrate how these integrated classes are mapped in application in one of the eleven pilot projects undertaken to test the provisional system of IEZ. The spatial extent of the integrated environmental zone VI (black – or unacceptably polluted for residential use) in Figure 13.1 is markedly smaller than is the same zone in Figure 13.2, since the combination of impact levels for each integrated class specified by Table 13.4 (existing conditions) is less demanding than for each of the same classes in Table 13.5 (new situations). Comparison of these two maps visually convey that the provisional system of IEZ provides less flexibility for locating new residential development than for allowing continuation of existing uses, unless action is taken to reduce the environmental impacts of manufacturing activities currently responsible for these spillovers.

Table 13.4   Existing situations classification of integral quality

Integral Class	Combination of sectoral classes
I_B	Only B, no C, D or E
II_B	2 x C, no D or E
III_B	At most 3 x C, no D or E
IV_B	At most 1 x D and at least 1 x C (or at most 4 x C, no D), no E
V_B	At most 2 x D (or at most 1 x D and 4 x C), no E
VI_B	3 x D or more, or 1 x E or more

Source: VROM 1990

Table 13.5   New situations classification of integral quality

Integral Class	Combination of sectoral classes
I_N	Only A, no B, C or D
II_N	2 x B, no C or D
III_N	At most 3 x B, no C or D
IV_N	At most 1 x C and at least 1 x B (or at most 4 x B, no C), no D
V_N	At most 2 x C (or at most 1 x C and 4 x B), no D
VI_N	3 x C or more, or 1 x D or more

Source: VROM 1990

Restrictions Associated with Integrated Classes

In addition to measuring and mapping the combined impacts of several forms of spillovers, the provisional system of IEZ specifies physical planning restrictions for each of the integrated environmental quality classes, ranging from no limitations for Class I to no residential uses permitted in areas mapped as Class VI. The range of restrictions for Classes II through V are based on the level of undesirability of the environmental load that each of these classes represent. Table 13.6 specifies the restrictions associated with each of the integrated classes for existing situations, and Table 13.7 for new situations. These restrictions are effect oriented, since they specify how areas with differing environmental quality may be used. Since reduction of pollution is expected to be undertaken first, success with this source oriented action will change the initial mapping of integrated classes as conditions surrounding sources of pollution improve.

Table 13.6   Physical planning restrictions applying to the integral environmental quality classes in existing situations

Integral Class	Physical planning restrictions
I_B	No restrictions
II_B	Projected and replacement residential construction allowed if there are no financially feasible alternatives available which are acceptable from a land use planning perspective or if the continuity of the neighbourhood is at risk
III_B	Idem Class II_B, but the local authorities should exercise greater restraint. Construction of projected housing allowed only to satisfy internal housing demand of municipality or residential centre in question
IV_B	Only replacement new construction allowed which does not significantly increase the population in the area
V_B	Only replacement new construction allowed which is indispensible to the continuity of the neighbourhood and which does not significantly increase the population in the area
VI_B	Removal of housing units

Source: VROM 1990

In summary, the procedure employed by the provisional system of IEZ consists of a series of steps. Initially, five categories of environmental impact from manufacturing are identified, and standards for acceptable and unacceptable levels of impact for each of these kinds of impact are prescribed. These standards serve as the end points for five classes of environmental impact (A through E) for each of the categories of spillover. These classes of impact are combined across the five categories of environmental spillover, and each of the resulting six integrated classes of environmental quality have associated with it a set of land use restrictions ranging from permitting residential uses in areas mapped as Class I, to prohibiting and even removing housing in areas designated as Class VI.

Table 13.7   Physical planning restrictions applying to the integral environmental quality classes in new situations

Integral Class	Physical planning restrictions
I_N	No restrictions
II_N	Residential construction allowed if there are no financially feasible alternatives available which are acceptable from a land use planning perspective
III_N	Idem Class II_N, but the local authorities should exercise greater restraint and only allow residential contruction if there are no acceptable alternatives available with an integral load of II_N or less. Construction allowed only to satisfy internal housing demand of municipality or residential centre in question
IV_N	New residential constrution allowed for housing urgently needed for land- or firm-specific reasons or for housing which fills in a gap in existing development and for which there is no acceptable alternative
V_N	New residential constrution allowed only for housing urgently needed for land- or firm-specific reasons
VI_N	New residential construction not allowed

Source: VROM 1990

13.3   Experimental Application of the Provisional System

Application of this program in the eleven pilot projects requires a procedure consisting of several steps. Pollution and hazard around manufacturing sites are measured and mapped, and compared with acceptable levels. Existing land uses surrounding these manufacturing sites are also mapped, primarily to identify residential areas which are considered as especially environmentally sensitive by the provisional system. Additionally, land use plans are reviewed to identify where new development is proposed. When heavily impacted areas do not include housing and are not intended for residential development, strict abatement measures for manufacturing firms are not necessary and industrial expansion may be possible. New residential development may be located at safe distances from sources of pollution. However, if housing is located in impacted areas or is proposed for these areas, extensive abatement may be required of manufacturing plants in order to assure acceptable environmental quality.

Employing this information, each local government must develop a sanitation plan identifying necessary abatement actions and addressing how to accommodate additional manufacturing activity without negatively impacting residential areas. In most cases, this plan will include both source oriented (abatement) and effect oriented (land use regulation) measures. Developing these requires a political process which accounts both for the financial and environmental implications of the plan. The provisional system requires that analysis of abatement consider whether 'it is possible and what it will cost to achieve an environmental quality that does not exceed the integral class IB at the most heavily loaded environmentally sensitive object' (VROM 1990, p. 21).

Current pollution control regulations in The Netherlands are based on best practicable means (BPM): 'those techniques which achieve the greatest reduction in emissions and which have acceptable costs for a normally profitable firm' (VROM 1990, Appendix 3.1). For Class VI areas remaining after BPM have been employed, the provisional system requires consideration of best technical means, as a way to convert black areas into grey areas, which would allow more flexibility in land use. Best technical means include 'those techniques which can achieve an even larger reduction in emissions than BPM and which have been applied at least once in practice, but which have higher costs than BPM (VROM 1990, Appendix 3.1). The provisional system stipulates that abatement is preferred to land use regulations, and that manufacturing firms will pay for the reduction of their pollution and hazard to acceptable levels.

In instances in which residential areas will still be impacted at unacceptable levels by environmental spillovers even if BTM are employed, either the polluting manufacturing activities would need to be shut down or impacted housing would need to be relocated. In many situations, emissions from several plants contribute to an unacceptable environmental loading on surrounding residential areas. Consequently it is difficult to determine how much each of these plants must improve their environmental performance in order to produce acceptable environmental quality. The provisional system recognises that these source oriented measures are not always the most appropriate action, and that effect oriented measures may be substituted. The land use regulations specified in Tables 13.6 and 13.7 are then employed. These regulations provide the local government with some alternatives, including accepting some non-negligible environmental impacts on residential areas, physically shielding housing from some effects, and in extreme cases relocating residents and removing housing from highly polluted areas.

Local decisions concerning the most appropriate set of source and effect oriented actions to take are incorporated into a long-term pollution management (sanitation) plan. This plan specifies pollution abatement required at the sources, and the final designation of the IEZ which specifies where housing is permitted in conformance with the land use regulations shown in Tables 13.6 and 13.7. This sanitation plan is included as an element of the land use plan of the local government, the mitigation provisions are monitored, and these provisions may be revised as changing conditions permit.

13.4 An Assessment of the Integrated Environmental Zoning Program

The IEZ program is an innovative and comprehensive approach to accounting for several kinds of environmental spillovers from manufacturing activity and, on the basis of mapping the spatial incidence of these spillovers, to seek pollution reduction or to limit residential use of areas which are heavily impacted. In this way, IEZ combines environmental management and land use planning.

The methodology for measuring and aggregating several environmental effects is explicit and well reasoned. Standards specifying environmentally acceptable and unacceptable conditions for residential areas establish the end points of a scale of performance for each kind of spillover included in the analysis. These standards are based on evidence concerning the health effects of exposure to these kinds of pollution and hazard, and on public opinion concerning what constitutes nuisance (de Boer, et al. 1991). In some cases, we have limited knowledge concerning the danger from various levels of some forms of pollution.

Aggregation of performance measures for several kinds of spillover is necessary if these effects are to be treated in a comprehensive manner. Since the adopted standards define the end points of the range from acceptable to maximum permissible levels of impact, the scales for each of the five categories of environmental impact are normalised, that is share a common scale. Each of these scales consist of classes of impact, meaning that absolute measurements are transposed into an interval scale, with some loss of information. Additionally, this simple combination of scores for various kinds of impact does not account for the ways in which various forms of pollution may interact to produce effects on health or quality of life which are greater than their sum.

As shown by Tables 13.4 and 13.5, the accumulation of scores across categories of pollution to produce an integrated environmental class (I through VI) for a location is not simply additive. For example, high scores for a few categories of impact result in the highest (most undesirable) integrated classes while lower scores for all categories result in more desirable integrated classes. This suggests that the summarisation technique used by the provisional system employs a weighing scheme, but the weights are not explicit or explained.

While the provisional system addresses only five categories of environmental effect, other forms of pollution may be added at a later stage, once people have more experience with this program and its application. Similarly, while the current program focuses on the impacts of manufacturing on residential activities, both the number of kinds of sources and of impacted land uses may be expanded in the future.

The eleven pilot applications of IEZ are already providing evidence concerning the applicability of this new program. Interviews of people involved in these pilot projects indicate considerable satisfaction with the provisional system, but in the judgement of many of these people the six integrated classes are too many and the prescriptions attached to each of these classes are too rigid (de Roo and van der Moolen 1991). The respondents would prefer more local discretion in determining the source and effect oriented measures to be taken. These decisions are difficult, since they involve costs to manufacturing firms and to local governments, and will have economic effects ranging from reducing the competitiveness of local firms to closing plants and losing employment (de Roo 1993). Determining a reasonable balance between these means of achieving acceptable environmental quality are thus political and differing situations suggest the desirability of allowing at least some local discretion in solving these problems. On the other hand, national specification of these requirements strengthens the position of local authorities in resolving difficult environmental issues, especially when there is local political opposition, and forces local governments to respond to rather than avoid environmental quality concerns.

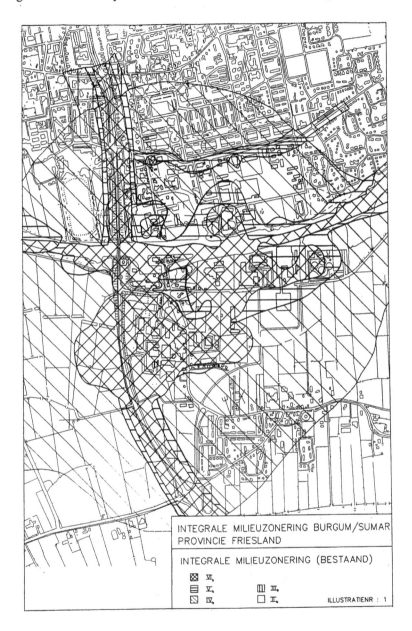

Figure 13.1 Burgum and Sumar existing situation
Source: Province Friesland 1991

13.5   Results from the Pilot Projects

The eleven pilot projects are distributed throughout The Netherlands, and range from an individual manufacturing site to a major industrial region. Other chapters in this volume address specific pilot projects in some depth and discuss in detail some of the difficulties which these projects have encountered. While not all of the pilot projects have been completed, preliminary results from several of them show that environmental quality in several cases is much worse than had been anticipated, and that the costs of successful mitigation in a number of instances will be greater than local public and private sector resources can support.

The pilot project in The Drechtsteden, a major industrial centre east of Rotterdam, found heavy pollution impacting six municipalities (Studygroup IMZS 1991). Major portions of Zwijndrecht, a city at the western edge of this region, have been mapped as IEZ Class VI (black), requiring closing manufacturing plants or relocating residents and removing housing. Much of the remainder of this city is mapped as Class IV, which under the provisional system permits only replacement residential construction with no increase in population. The other neighbouring cities are similarly but less heavily impacted. Since this region has a large population which is heavily dependent on manufacturing employment, compliance with the requirements of the provisional system will be very difficult.

Similarly, a rendering plant between the towns of Burgum and Sumar places the latter and parts of the former in the integrated class which prohibits new residential construction, as illustrated in Figure 13.2. Sumar has postponed developing housing in the impacted area, although it has been designated for this use in the land use plan and the municipality wants to accommodate additional population. Figure 13.1 demonstrates that less demanding requirements for existing conditions than for new development results in spatially smaller mappings of the environmentally least desirable integrated classes. Finally, the pilot project in Hengelo found that an extensive area is impacted by a combination of air pollution, odour, and industrial hazard. Consequently, some residential areas are designated for demolition and location of new housing will not be permitted in a large portion of the city.

13.6   Conclusions

As these illustrations point up, environmental conditions in several of the pilot project areas are considerably worse than had been expected, and remediation is more expensive and difficult than anticipated. These cases also demonstrate the value of the IEZ methodology for assessing environmental quality in residential areas, as well as the desirability of providing localities with some flexibility in designing workable solutions to the environmental problems that are identified by employing this methodology.

Environmental quality standards employed by the provisional system of IEZ account both for threats to health and to well being, and attempt to represent levels of public tolerance for pollution. As with other public programs, there may be wide agreement on what is sought while disagreement over means of achieving these results. The Dutch program of Integrated Environmental Zoning is a bold and comprehensive initiative to measure and assess the environmental impacts of manufacturing activities on surrounding residential areas and to prescribe means to achieve desirable environmental quality. The design of this program and its application in a set of pilot projects provides valuable lessons in methodology and implementation which can be adapted by other countries to address increasingly recognised and urgent environmental concerns in urban areas.

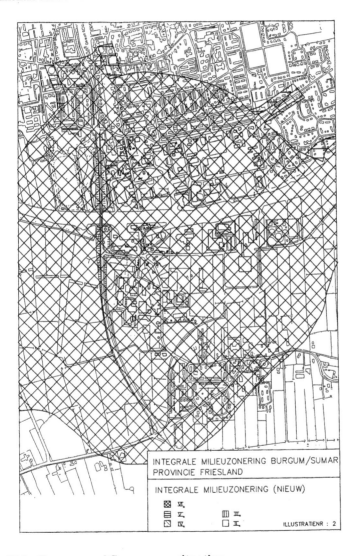

Figure 13.2   Burgum and Sumar new situation
Source: Province Friesland 1991

Note

1   Donald Miller is Professor at the Department of Urban Design and Planning, University of Washington, Seattle, USA.

References

Boer, J. de, H. Aiking, E. Lammers, V. Sol and J. Feestra (1991) Contours of an Integrated Environmental Index for Application in Land-Use Zoning, in O. Kuik and H. Verbriggen (eds) *Search of Indicators of Sustainable Development*, Kluwer Academic Publishers, Dordrecht, The Netherlands.

Province Friesland (1991) Integrale Milieuzonering Burgum/Sumar, Tussenrapport, Hoofdgroep Waterstaat en Milieu, Bureau Geluid en Lucht, Leeuwarden, The Netherlands.

Roo, G. de, and B. van der Moolen (1991) *De Voorlopige Systematiek voor Intregrale Milieuzonering, een Doelgroepbenadering in Drie Proefprojekten*, Geo Pers, Groningen.

Roo, G. de (1993) Environmental Zoning: The Dutch Struggle Towards Integration, *European Planning Studies*, 3, pp. 367-377.

Studygroup IMZS-Drechtsteden. (1992) *Inventarisatie Milieubelasting IMZS-Drechtsteden, Eerste Rapportage* [Inventory Environmental Hindrence IMZS-Drechtsteden, Report of Phase I], The Hague.

VROM [Dutch Ministry of Housing, Spatial Planning and Environment] (1989a) *Nationaal Milieubeleidsplan* [National Environmental Policy Plan], Second Chamber Session 1988-1989, 21137, No. 1-2, The Hague.

VROM [Dutch Ministry of Housing, Spatial Planning and Environment] (1989b) *Integrale Milieuzonering deel 2: Projectprogramma cumulatie van bronnen en integrale milieuzonering*, VROM, Leidschendam, The Netherlands.

VROM [Dutch Ministry of Housing, Spatial Planning and Environment] (1990) *Ministerial Manual for a Provisional System of Integral Environmental Zoning*, IEZ-report No. 14, The Hague.

VROM [Dutch Ministry of Housing, Spatial Planning and Environment] (1991a) *The Integration of Environmental Legislation in The Netherlands*, VROM, Liedschendam, The Netherlands.

VROM [Dutch Ministry of Housing, Spatial Planning and Environment] (1991b) *National Environmental Policy Plan*, SDU Publishers, The Hague.

VROM [Dutch Ministry of Housing, Spatial Planning and Environment] (1992) *Environmental Policy in The Netherlands*, VROM, The Hague.

Chapter 14

The Rise and Fall of the Environmental Zone: A Discussion About Area Oriented Environmental Planning in Urban Areas

G. de Roo[1]

14.1 Introduction

In The Netherlands, differing land uses are organised by means of zoning. Within these spatial zones, certain functions are permitted and others are not allowed. Consequently, there is much experience with this instrument of spatial planning. The latest development is called environmental zoning. Zones are being used to separate environmentally intrusive activities and environmentally sensitive areas in a sustainable manner. Environmental zones indicate the areas in which some forms and levels of environmental pollution are acceptable: chiefly manufacturing areas. Housing may not be located in an industrial area, whereas industry must not environmentally impact areas that are meant for residential functions.

The principle of zoning is old, but the application of this instrument to multiple forms of environmental impact poses unexpectedly large, interesting and fundamental problems. Thus this chapter discusses mapping impacted areas using a classification system which combines measurements of several kinds of pollution, and problems arising from current interest in mixing urban activities in developing compact urban centres.

14.2 Environmental Policy versus Spatial Planning

Environmental policy has two lines of approach to prevent problems in urban areas. On the one hand the policy aims at the different economic activities that may generate unacceptable environmental impacts. This type of policy tries to reduce the spatial range of pollution by means of decreasing emissions at the source (VROM 1988-1989).

The second approach is area policy, which focuses on measuring the environmental quality found in areas surrounding sources of pollution, and regulating the kinds of environmentally sensitive activities such as residences that may be located in these polluted areas. This policy is interesting for two reasons. The first reason is that only recently have we realised that many residential and other areas are dangerously polluted. Abatement policy is therefore necessary in order to improve existing conditions in an area (VROM 1990a). A second reason is to prevent increased pollution, a policy which will also have to be maintained in the long term. That is, it will remain necessary that different area functions are sustainably separated from each other. Houses are no longer to be built in or next to industrial areas. Noisy roads must not be located where they can disturb residential areas with traffic noise. Agricultural areas can no longer be developed in or near natural areas (VROM 1989). This form of area-focused policy requires not only environmental management for pollution reduction but also spatial planning, in order to provide long-term environmental quality.

The desirability of combining environmental management and spatial planning is rather obvious, since both fields share an important main objective. Both aim at improving and maintaining the quality of the environment or the quality of life. Unfortunately, as yet this objective is one of the few agreements between these two lines of policy. The ways in which each strives to meet its goal lie miles apart (VROM 1990a).

14.3   Mixing Activities versus Separating Activities

Spatial planning aims at a good social environment mainly through interrelating urban functions; making them accessible to one another. Anywhere people are residing, children need to go to school as well, and people must be able to shop and go to work. The same residential location must be properly accessible by the road-system and the public transport. Environmental policy seeks to achieve its goal by sustainably separating environmentally sensitive areas and environmentally intrusive activities. Therefore, environmental policy strives after a separation of functions.

The search for an optimum between spatial integration and separation is for an important part carried out by means of the instrument of environmental zoning. In principle, environmental zoning ought to be the instrument by which both environmental policy and spatial planning determine the quality of the social environment. Unfortunately, the effect of the environmental policy content within spatial planning is largely limited to the translation of standards in the local land use plan. Environmental policy has become prescriptive for spatial planning.

There is little discussion of co-ordinating and balancing these two lines of policy. Spatial planning is often legally confined to dealing with the location of land uses, based in part on environmental standards. However, when environmental standards become translated into environmental zones, areas identified as having low environmental quality are often so large that, if only a few land uses are permitted within them, spatial planning to accommodate many other activities is severely limited. Ways must be devised so that both policy fields can make their contributions to improving environmental quality, and to make tradeoffs between them (de Roo 1993).

14.4   Environmental Zoning

The first type of environmental zoning deals with noise nuisance. Along motorways and railroads and around airports and industrial areas much effort is being made to map out the noise load. In fact this has been done for about ten years now, and quite successfully. When the noise load is mapped along motorways and railroads, the locations can be found where the noise load must be considered too high. When the zone which represents the environmental overload appears to contain houses as well, measures have to be taken. These measures can be taken at the source, which will be very difficult in the case of traffic noise. Apart from that, measures at the impacted land use can reduce the noise load, for instance through housing insulation. In The Netherlands, a method that is often used is the installation of sound-proof barriers which are built between the road and the environmentally sensitive area, that is measures between source and recipient (Noise Abatement Act 1979). The way noise pollution has to be dealt with in The Netherlands is effectively regulated in the Noise Abatement Act. Noise standards, zones and the planning consequences linked to these have been summarised in the act, and much work is being done on the enforcement of this act. Many sound-proof barriers are being built, as well as programs for housing insulation, the result of which is that the problem of noise pollution will largely be eliminated in the years to come.

There are other environmental problems which can in principle be solved by zoning as well. These environmental problems have to do with, among other things, odour, vibration, dust, and the emission into the air of toxic and carcinogenic substances. These forms of environmental load can be roughly divided into roughly three categories. The first is nuisance, and includes noise, odour, vibration, and dust. Characteristic of these forms of nuisance is they do not threaten life, but can cause a good deal of irritation and complaints. The standard for nuisance is therefore based upon the percentage of a population who are impacted. Present regulations are only sufficient for noise. The prescription and the zoning of other forms of nuisance is still being done largely on the basis of directives, although much is being done to improve clarity (VROM 1990b).

Another category is risk, which may be life threatening in acute situations. Industrial calamities such as fire and explosion, emission of toxic and carcinogenic substances into the air are major forms of risk. The risk standard is expressed in terms of the number of deaths per annum as a result of exposure to each level of environmental impact. These standards have been considered to be very demanding, and new guidelines are being developed (Dutch Emission Guidelines 1992).

These two categories are especially important for human beings. Zoning on the basis of these categories is therefore of particular significance in urban areas. A third form of environmental impact which can be assessed and zoned is injurious effects for the ecosystem. In The Netherlands, this form of zoning aims mainly at the emission of ammonia from farming (Ministry of Agriculture, Nature and Fishery 1992-1993). The standard is expressed in terms of the deposition of an amount of acid equivalents to the hectare per annum. Zoning on the basis of ammonia has been obligatory for some time now, but leads to heated discussions between farmers, eco-groups and authorities. The result is that the standard has come under a considerable attack, although the establishment of standards for ammonia are specified in legislation.

Zones which reflect the environmental impact for noise are mainly localised. The load often reaches no further than a few hundred meters, which makes dealing with the noise problem manageable. This is not the case with odour and toxic and carcinogenic substances in the air, resulting from some industrial production processes. Acute forms of environmental impact for these effects may sometimes have a size of several square kilometers. Consequently, the affected area reaches well into residential districts and to other environmentally sensitive activities. In these instances, only measures taken at the source, thus at the industrial plant itself, are effective. If adequate means of cleaning up emissions are not possible, the only alternative may be closing the factory, with the unwelcomed loss of jobs (de Roo 1992).

14.5   Environmental Standards

The Dutch environmental policy includes a system of standards which takes industry into account. The object of this system is that the economic production process does not have to suffer excessive damage, while environmental standards that are considered desirable must be met within manageable time (within one generation). The Dutch system of standards is founded on the concept of limits, targets and interim targets. The limit states the level above which the environmental impact is considered unacceptable. The target states the level below which the environment suffers negligible effects. This interim target expresses the immediate environmental standard, somewhere between the limit and the target, and is modified periodically. This means that the interim target moves one step closer to the target every four or five years. In this way the target, which is the value which shows high environmental quality, can be reached within twenty-five years. The economic production process must improve its performance through time, and when replacement investments are being made, the cleanest methods can be taken into account (VROM 1990-1991).

The system of standards with two extreme values, namely the limit and the target, with an interim value that moves from limit to target, is called progressive standard-setting. Despite the progressive standard setting, it appears to be extremely difficult for many industries to meet the target. Industrial firms are not always to blame for this, since the government has at times been indecisive and careless in the issuing licenses to industries.

14.6   Integral Environmental Zoning

Despite the problems with the different forms of environmental zoning, in The Netherlands experience is being acquired with a new type of zoning, the integrated environmental zoning (VROM 1990b). Considering the problems already mentioned it is not surprising that this new form of zoning has caused heated debates. Nevertheless friend and foe consider it sensible to figure out the combined environmental impact of several forms of pollution rather than regulate them one at a time. A sectoral approach to environmental problems in a complicated industrial area has a disadvantage that successful reduction of one form of pollution may increase another form of pollution, giving rise to new forms of regulation, which increases the uncertainty for manufacturing firms. When this happens, firms are forced to adapt their production process time and again, but each time on the basis of different requirements. Both the authorities and industrial circles want clarity about environmental quality requirements, which is the promise of the integrated method of environmental zoning (de Roo 1992).

What does integrated environmental zoning entail, and why is it controversial? Integrated environmental zoning seeks to separate sustainably environmentally intrusive activities and environmentally sensitive areas on the basis of the complete set of environmental impacts caused by industrial activities. This means that noise, odour, emission into the air of toxic and carcinogenic substances, and other forms of spillovers are put together in order to determine the complete environmental impact.

A method has been developed with which these different forms of environmental effect can be added up. This is done on the basis of a classification of spatial units or analysis zones. When an area falls into the lowest category this means 'free from impact', and appears as white on a map of environmental impact. However, when an area is given the colour black on the map, this means that for this area an inadmissible environmental load has been measured in one of the sectoral forms of environmental impact. Thus, when one of the sectoral forms of load exceeds the maximum standard value, the ceiling has also been reached for the complete environmental load in that area. When the upper limit has been exceeded, either pollution reduction must occur or sensitive activities that are unacceptably impacted must be moved. Measures at the source are preferred. Between the black and the white designations, some shades of grey are used as well, which indicate undesirable but not unacceptable levels of impact, each of which have planning consequences (VROM 1990b).

An advantage of this method is that it clearly indicates whether an environmentally sensitive area is experiencing an inadmissible level of environmental impact. Also corrective measures can be connected to each class of environmental load. These corrective measures depend on the class and hence on the colour on the map into which an area falls.

Integrated environmental zoning additionally provides a means of combining the effects of several forms of pollution and hazard. For example, the amount of noise pollution and the amount of nuisance caused by odour are taken together, so that the complete load or impact becomes prescriptive for the area.

14.7 Black Areas

In some of the largest Dutch industrial areas, many forms of environmental impact occur, affecting large portions of cities. When the sectoral environmental impacts are combined, it is often the case that only a few so-called white areas remain on the map.

The procedure described here is how to get at a map that indicates which areas are suffering from environmental load and in which amount. This procedure has been carried out in a number of test cases in The Netherlands, and the results are beyond expectation. Large parts of urban areas are designated as having unacceptable environmental quality, based on the standards employed. The consequence of this is that a very severe abatement will have to be carried out at the source (VROM 1990b), but if this is not sufficient, even houses will have to be pulled down and residents relocated (ten Cate 1993).

This integrated approach is not yet based in legislation. About twelve test projects are underway in which the combined effects of environmental impacts are being mapped. Abatement agreements are being sought on the basis of these data. In some cases, where adequate abatement is not possible, residential areas are being relocated. However agreements have not yet been reached in the most controversial cases. In three of the twelve test projects, the combined environmental impacts are so great that only two options are available if environmental quality is to be brought back to the level considered desirable. These two are pulling down thousands of houses, or closing down the most intrusive factories. Both options are socially undesirable and very difficult for the public to support. This situation reveals a major problem with the approach used in the pilot projects: the lack of flexibility for finding acceptable solutions to unacceptable environmental conditions.

14.8   Flexibility

Apart from the twelve test projects, it is estimated that there are another 250 'complicated' industrial areas in The Netherlands (VROM 1991). The companies established in these areas cause such a complicated set of environmental impacts that integrated environmental zoning in its present form will likely pose equally difficult decisions concerning what must be done to mitigate these conditions. Consequently, pressure is increasing to relax the proposed regulations regarding environmental demands (Kuypers 1993), and the Ministry of Environment now realises that the provisional approach will lead to unexpectedly large problems. Therefore the Ministry is considering allowing flexibility in the regulations for especially difficult cases, such as setting the interim target higher when there are good reasons to do so (Wolters 1993). When the integrated environmental zoning system would result in pulling down thousands of houses, such flexibility will be seen as justified.

Critics believe that flexibility in the setting of standards will assist in providing a healthy social environment. Yet they urge as well flexibility in time and differentiation of area as ways to speed up the integration of spatial and environmental policy (de Roo 1993; Peters 1993; Voerknecht 1993). Flexibility in time means that when a proper solution to an environmental overload is not possible in the short run, then more time should be taken to arrive at the situation desired. This will delay improvement of undesirable conditions, but avoid drastic actions which have comparably undesirable effects in terms of dislocations and job loss.

Despite large black spots which indicate areas on a map with an unacceptably high environmental load, environmental zoning proves to be a sensible instrument. Thanks to an inventory the (complete) environmental load in an area is made perceptible, and a basis is established on which it is possible to work at a durable solution of an area's environmental problems.

Although environmental policy in The Netherlands is rapidly decentralising, it is still largely aiming at the environmental standard, which thus becomes the standard framework for other forms of policy, among which is spatial planning. This is a pity, because practice shows that when environmental demands are looked at too much, then very little remains possible in a certain area. The other proposal, area oriented differentiation, calls for differing standard setting for important urban areas, nature reserves and remaining areas. Current standards would apply to 'remaining areas', whereas stricter standards would apply in the case of nature reserves. Greater flexibility would be used in setting standards for important urban areas (Dorren 1993). Reasoned and justifiable flexibility, in time and by area, hold promise as means to developing a politically acceptable adaptation of the provisional system of integrated environmental zoning; one that can be established by national legislation.

14.9   Conclusion

Despite alarming findings from the pilot projects, that larger than expected areas are classified as of unacceptable environmental quality, integrated environmental zoning proves to be a sensible instrument. The weaknesses of inflexible standards have been identified, as has the need to trade off between environmental quality and economic activity, at least over the short run. Similarly, the need to integrate spatial planning and environmental management is coming to be recognised as a means of improving both environmental quality and the social environment. A continuing agenda will be necessary to find ways of fostering compact urban development, which has been an objective of spatial planning, while protecting the health and well-being of inhabitants of cities as conflicting activities are moved yet closer to one another.

Note

1   Gert de Roo is Professor in Planning at the Faculty of Spatial Sciences, University of Groningen, Groningen, The Netherlands.

References

Cate, F. ten (1993) Consequenties van 'onwrikbare' milieuzones niet meer te accepteren, *Binnenlands Bestuur*, Vol. 14, pp. 23-25.
Commissie Emissies Lucht (1992) *Dutch Emission Guidelines*, Stafbureau NER, Bilthoven, The Netherlands.
Dorren, G. (1993) Jan Cleij geïrriteerd over milieuregels en uitzonderingen, *ROM*, Vol. 5, pp. 3-5.

Kuypers, C.J. (1993) De juridische kaders van milieuzonering, in G. de Roo (ed.) *Saneren en Bestemmen*, Rijksuniversiteit Groningen, Groningen, The Netherlands.

Ministry of Agriculture, Nature and Fisheries (1992-1993) *Tijdelijke regeling inzake de ammoniakdepositie veroorzaakt door veehouderijen* (interimwet ammoniak en veehouderij), Kamerstukken II, 23221, No. 1-3.

Noise Abatement Act (1979) Stb. 99, SDU, The Hague.

Peters, G.H.J., and J. Westerdiep (1993) Kop van Zuid, in G. de Roo (ed.) *Kwaliteit van norm en zone*, Geo Pers, Groningen, The Netherlands.

Roo, G. de (1993) Patstelling of Synthese?!, in G. de Roo (ed.) *Kwaliteit van norm en zone*, Geo Pers, Groningen, The Netherlands.

Roo, G. de (1992) Milieuzonering stuit op planologische obstakels, *ROM*, Vol. 6, pp. 14-17.

Roo, G. de (1993) Zonering: integratie tussen milieu en ruimtelijke ordening, *ROM*, Vol. 6, pp. 21-24.

Roo, G. de, and B. van der Moolen (1991) *De Voorlopige Systematiek voor Integrale Milieuzonering*, Geo Pers, Groningen, The Netherlands.

Voerknecht, H.C. (1993) De Drechtsteden, in G. de Roo (ed.) *Saneren en Bestemmen*, Rijksuniversiteit Groningen, Groningen, The Netherlands.

VROM [Ministry of Housing, Physical Planning and Environment] (1988-1989) *National Environmental Policy Plan*, Kamerstukken II, 21137, No. 1-2.

VROM [Ministry of Housing, Physical Planning and Environment] (1989) *Integrale milieuzonering deel 2: Projectprogramma cumulatie van bronnen en integrale milieuzonering*, Leidschendam, The Netherlands.

VROM [Ministry of Housing, Physical Planning and Environment] (1990a) *Plan van Aanpak voor het gebiedsgericht milieubeleid*, SDU, The Hague.

VROM [Ministry of Housing, Physical Planning and Environment] (1990b), *Integrale milieuzonering deel 14: Ministerial manual for a provisional system of integrated environmental management*, Leidschendam, The Netherlands.

VROM [Minstery of Housing, Physical Planning and Environment] (1990-1991) *Milieukwaliteitsdoelstellingen bodem en water*, Kamerstukken II, 21990, No. 1.

VROM [Minstery of Housing, Physical Planning and Environment] (1992) *Integrale Milieuzonering deel 23, Omvang zoneerbare milieubelasting 1991*, The Hague.

Wolters, G.J.R. (1993) Visie van de Rijksoverheid, in G. de Roo (ed.) *Kwaliteit van norm en zone*, Geo Pers, Groningen, The Netherlands.

Chapter 15

Integrated Environmental Zoning in the IJmond-Region near Amsterdam

D. Arbouw[1]

15.1 Introduction

In the 1980s the province of Noord-Holland, the municipalities of Velsen, Beverwijk and Heemskerk (Figure 15.1) and local industry agreed to search for a method of integrating measures of several forms of pollution into an assessment of the total environmental load in the region. A steering committee was formed, and a study was made, resulting in a report called 'Experiment of integration of environmental load in the 'IJmond region' (IVM 1988). When the national government proposed to participate in an experiment on national scale, the steering committee agreed (VROM 1990).

This chapter discusses the history of this approach, the system of integration and the results of the studies. Finally we discuss the expected agreement in solving the remaining environmental load by abatement at the sources and by spatial planning in the region.

15.2 General Situation

The Netherlands borders the North Sea and is also known as the gateway to Europe. (Figure 15.1). The Netherlands, in comparison with other European countries, has a high population density, with an average of more than 400 people per square kilometre, and in some parts of the country as high as 1,000 people per square kilometre. The highest density areas are in the western part of the country which is dominated by four cities; Amsterdam, The Hague, Rotterdam and Utrecht, and 35 per cent of the total population of the country lives in these cities.

Almost the entire country is cultivated in one way or another. Even the nature reserves, mostly woods, have been laid out by man. Furthermore The Netherlands is a highly industrialised country. Town and country are influenced by this. The result is that many different interest groups and coalitions are in competition for space which is scarce. This issue has to be handled carefully. Town planning is often a politically argumentative issue.

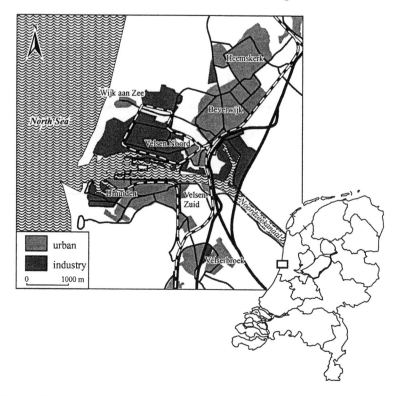

Figure 15.1   Local situation of the municipalities of Velsen, Beverwijk and Heemskerk

An example of this conflict between economic, environment and physical planning is the subject of this chapter. This project was started as a contribution to the collective responsibility for land-use planning by the province, municipalities and industry. In the beginning, we expected that these efforts to reduce environmental load in the eighties were sufficient, and that this project was going to prove this to be the case.

15.3   Local Situation

The enormous steelfactory 'Corus' (formerly known as 'Hoogovens') is the subject of this project. It produces about six million tons of raw steel a year, and employs 11,500 workers (Hoogovens 1993). The participants in this project feel a mutual responsibility for the environmental situation and the use of the land around the steel factory, which has grown in last four decades. (Figure 15.1).

When this steelplant was founded in 1918, it was situated at a seaport, so the infrastructure for raw materials was and still is ideal. Nowadays 7.5 million tons of iron and 3.7 million tons of coal are imported annually (Hoogovens 1993). In the 1920s, in the neighbourhood of the steel factory there was no residential area. When the works expanded, it was necessary to build residential areas for the workmen. It became a region which needed planning for the use of the land and licence regulations for the large furnaces of the factory. In the 1970s and 1980s, it became necessary for the region to seek ways for reducing the environmental load of noise, odour or smell, and dust (Provincie Noord-Holland 1981).

Noise, particularly impulse and tonal components, were held to constitute most of the nuisance. Projects were started to reduce the sound, and noise barriers were built. To reduce dust, techniques included water spraying systems on the stockyards, by road cleaning, and closed systems for coke ovens and blast furnaces. Bad smell or odour, produced by coke ovens and cinder granulation, also needed to be reduced. Measures taken to solve these problems were, among others, closed systems, gas cleaning and slag granulation with water and a high chimney.

Today in The Netherlands, we have an Environmental Control Act (1993) which contains new targets for environmental reductions on local level; some of them to contribute to a world-wide reduction. Examples of the chapters in this act are:

- *air pollution*: reduction of Sulphur dioxide and oxides of nitrogen;
- *heavy metals, and dioxides*: as toxic materials to be reduced;
- *climatological effects by changing energy consumption soil protection and reduction of waste and dust to prevent the urban area from environmental damage.*

The steel factory Corus pays about 110 million dollars a year for taking reduction measures (Hoogovens 1993). To prevent new problems, the steelfactory urges a restriction of environmentally sensitive new activities near this industrial area.

15.4   Project

This project, integrated environmental zoning, started with this goal:

> Tuning of decisions in case of environmentally intrusive activities and environmental sensitive activities to achieve distance and/or reduction of industrial emissions (Provincie Noord-Holland 1991).

Participants in the project are the province of Noord-Holland, the local municipalities and the industry. Funding is primarily from the National Government.

15.5   Studies

Two studies have been made to get a clear view of the environmental load in the region: top down, focusing on zoning based on technical knowledge and measures; and bottom up, with emphasis on an inquiry concerning nuisance in the neighbourhood (what is the meaning of this to the inhabitants in this area).

Top-Down Technical Study

We have determinated the emissions in the air from every significant plant for odour, carcinogenic and toxic components, and noise. The emissions of fine dust have been measured for many years (Provincie Noord-Holland 1992). Based on this evidence, we could calculate the emissions of odour, noise, and fine dust in the study-area.

Based on locally desired or agreed upon target values, interim target values and limit values, we have built four environmental quality classifications. The zones, based on these values, are printed by the BIMZ system (VROM 1991), a computer aided Geographical Information System. Comparable classifications for these different factors allow comparison of pollution conditions.

Odour measured in odour units (1 ou is defined as 50 per cent of a panel can still smell it) is based on Hydrogen Sulphide, being the most specific recognised odour for Corus and probably accounts for most emitted odour. The quality classifications are based on proposed legislation (VROM 1991). Nearly the whole area is covered by an environmental load of odour (Figure 15.2) (TNO 1991, 1992, 1992).

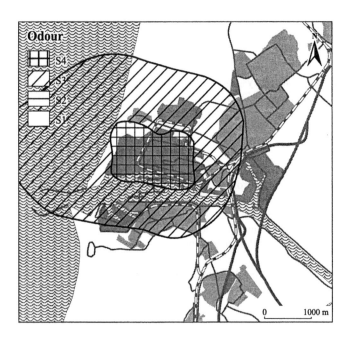

Figure 15.2   Odour caused by Corus in the IJmond-region

Noise classifications are Figure 15.3, noise based on industrial and traffic noise. The quality classifications are based on the values in the noise abatement act (VROM 1978). These zones of impact are smaller in scale (Figure 15.3) (M&P 1991).

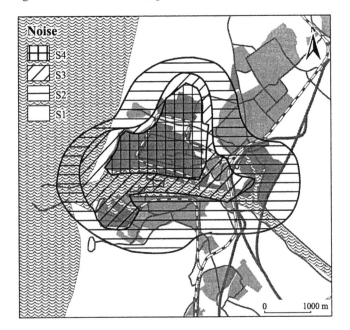

Figure 15.3   Industrial and traffic noise in the IJmond-region

Fine dust zones are produced by first searching for a virtual emitting point based on the emissions around the industrial area. In Figure 15.4 is a zone calculated by a diffusion model. The quality classifications are based on the PM 10 standard (Federal Register 1987). The zone of acute impact from fine dust is relatively small (Provincie Noord-Holland 1992).

Major hazard, carcinogenic and toxic components could not be printed on this scale since they affect a small area at most. The effects caused by transport of gas, petrol and dangerous chemicals are of local interest. Major hazard from industry remains within the industrial area (SAVE 1991; TNO 1991).

Using this methodology, the study leads to the clear definition of zones, based on the chosen classification system. The results could be a theoretical basis for physical planning. Restrictions, often with uncertain validation, are however the basic means of improving environmental quality, along with arithmetical translation of measurement data into zones.

Bottom Up

There has been made an inquiry concerning nuisance from odour, noise, dust, and the impact of these on satisfaction of living in the residential areas. The inquiry in the neighbourhood with 125.000 inhabitants included 1,200 participants, and employed a standard survey by phone (VROM 1991). The results provide the following information (OP&P 1993).

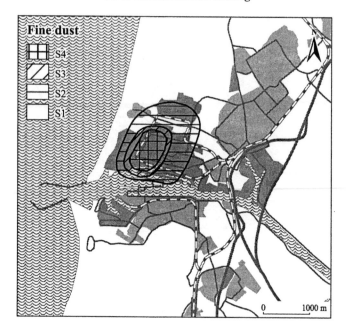

Figure 15.4   Fine dust zones in the IJmond-region

Figure 15.5 shows the data concerning the nuisance effects of noise, odour and dust related to distance. Most nuisance is given by dust, next odour and then noise. Nuisance in relation to distance (from a virtual centre of the industrial area) does not diminish after 4 kilometres.

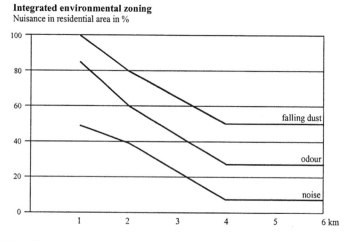

Figure 15.5   Nuisance in relation to distance

Information was also developed concerning the desirability of integrated environmental zoning. Satisfaction drops from 62 per cent to 51 per cent based on nuisance, when people are asked to consider a single (S/Sectoral) or several (I/Integrated) environmental loading factors or forms of pollution. Based on severe nuisance, satisfaction drops from 57 per cent to 34 per cent when considering only one (S/Sectoral) versus several (I/Integrated) environmental factors (Figure 15.6).

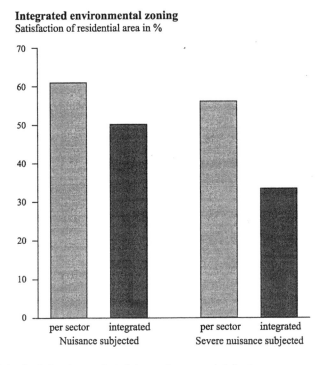

Integrated environmental zoning
Satisfaction of residential area in %

Figure 15.6   Satisfaction, related to environmental factors

The data in Figures 15.5 and 15.6 are useful information for physical planning and fundamental support the necessity of pollution reduction at the source. Table 15.1 shows the relation between local and national nuisance (OP&P 1993).

Measures taken in the 1970s and 1980s were sufficient for improving environmental quality with respect to several forms of pollution, as noted earlier. These are conclusions we expected:

- industrial noise is relevant, and the contribution cannot be ignored, as it is laid down in legislation;
- legislation and firm restrictions prevent major hazard in the residential area. Major hazard remains only near busy transport roads. Also carcinogenic and toxic components are not relevant on this scale;

- the levels of satisfaction in general with the residential area does not differ from the national situation.

Table 15.1   **Environmental nuisance in percentages**

	IJmond-Region	The Netherlands
Dust	63	-
Industrial odour	40	18
Traffic odour	9	8
Traffic noise	21	34
Industrial noise	16	6
Neighbourhood disturbance	25	27

Our expectations appear to have been confirmed. The unexpected conclusions include that fine dust is not a problem, as a result of the work done in last years. Falling dust however does result in severe nuisance in the neighbourhood. Besides, a study of the relation between dust and nuisance in 1991 shows the possible effect of falling dust with particles > 10 µg and nuisance (LUW 1991). Neither of the effects of odour on greatly reduced sense of environmental quality were expected. Restrictions for physical planning of residential areas based on odour classifications will be a great problem, since there is no legislation on which to base what action should be taken.

	Classi-fication	Industry	Residential area
Limit value	I4 S4	reduction emissions	sanitation
	I3	short term	fewer occupants
Optional value		REDUCTION EMISSIONS	NO NEW HOUSES
	S3	long term	equal occupants
	I2	reduction emissions based on nuisance	exemption
Target value	I1	--	exemption
	S1	--	--

Figure 15.7   **Restrictions based on sectoral and integral classifications**

Both the study of zones based on measured pollution levels and the survey proved to be necessary and supplement one another. Critical contributing factors to reduced environmental quality stemming from noise, falling dust and odour are relevant for physical planning. However, there is no measurement zone for falling dust as no reliable data are available. Falling dust will be a new area for analysis in which various forms of dust (such as beach and dune sand or traffic dust and industrial dust) will be assessed for their impacts.

15.6   Integration

The necessity of integrating measures of several forms of pollution is proven. Multiple environmental factors reduce the satisfaction of people with environmental quality significantly, as shown in Figure 15.6. Figure 15.7 illustrates the chosen system of integration in this project. It is a quite simple method of summing the environmental loads from several contributing factors. When one factor is relevant, it is called a sectoral (S) classification. When there are two equal relevant components (or factors), the classification is called an integrated (I) classification, with more restrictions. The higher the classification, the more restrictions. The restrictions are both for industry and for residential area in case of physical planning. This does not depend on any GIS system, however a GIS may be employed. This classification of conditions may be done by laying maps on maps. Based on locally chosen classifications, summation of every kind of environmental load can be made.

15.7   Conclusions and Solutions

Figure 15.8 shows present situation, integrating the contributing factors of noise and odour according the system described earlier. It was also noted earlier that physical planning is often a source of political contention. Figure 15.8 illustrates how this contention emerges, especially when existing conditions are related to the restrictions set forth in Figure 15.7.

Severe restrictions to uses permitted in physical planning for areas heavily impacted (class 3) greatly limit development in much of the urban area. No new housing may be built within the space with sectoral classification 3 (S3), nor with integrated classification 3 (I3). The political question soon will be if and when industry will reduce its environmental emissions, assuming that such reduction is possible. Environmental problems resulting from several contributing factors are too difficult to be solved in the short term.

Short-term efforts to improve environmental quality in the impacted area include trying to solve the problem of odour by finding the emitting plants and requiring them to reduce their pollution. This may include replacement or outplacement of specific parts of factories in future, or emission reduction in the short term. This research started some time ago (DHV 1993).

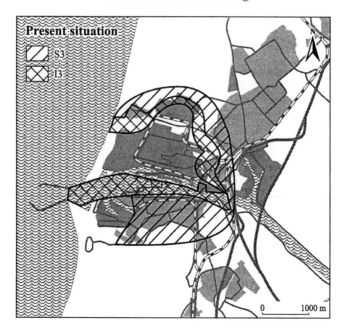

Figure 15.8   Present situation in the IJmond-region

Industry is already committed to reduce dust emissions by some 50 per cent, from 1985 levels. The goal was to achieve this in 1995 (NMP+ 1991). Research for ways to make greater reductions will be necessary. The political issue in this instance involves how certain we can be about levels of future dust reduction, and thus whether areas that are now impacted will be clean enough in the future to accommodate new residential areas. The final situation (Figure 15.9) is based on abatement of odour and noise. The goal is no more environmental load in most residential areas higher than a sectoral classification 2 of every contributing factor. If this happens, new housing can be built in restricted quantities in integrated classification 2.

For long term planning, industry must develop an 'environmental management system' for improving its environmental performance. A regional plan will be made by the province, with an integrated environmental zone, which will be used as the basis for licence regulations for industry, and as the basis for exemptions for building houses in areas where pollution will be acceptably reduced. Municipalities will produce local land-use plans in this area, based on the regional plan and the sectoral and integrated environmental classification scheme. We hope that this project will make a contribution in finding ways to tackle the environmental load problem.

The integrated environmental zoning project in the IJmond-region, located near Amsterdam, has provided a clear assessment of pollution impacts on this area, and their implications for development. This combination of abatement and restrictions on development will lead to a solution of the conflict between industrial expansion and the need for new residential areas. Solving the problems which this project discovered requires the collaboration of all of the parties, both those generating the impacts and those being impacted. Collective responsibility proves to be the key in a project like this. We have confidence that this collective responsibility will be forthcoming, and effective.

Figure 15.9   Final situation in the IJmond-region

Note

1   Dirk Arbouw is Senior Advisor Environmental Affairs in the Province Noord-Holland, Haarlem, The Netherlands.

References

DHV (1994) *Integrale Milieuzonering IJmond, Geur en H₂S*, Dienst Milieu en Water Provincie Noord-Holland, Haarlem, The Netherlands.

Federal Register (1987) *Revisions to the National Ambient Air Quality Standards for Particulate Matter*, Environmental Protection Agency, Washington, DC.

Hoogovens (1993) *Milieu jaarverslag van het staalbedrijf Hoogovens IJmuiden*, IJmuiden, The Netherlands.

IVM (Instituut voor Milieuvraagstukken) (1988) *Experiment integratie milieubeleid IJmond*, March 1988, Vrije Universiteit, Amsterdam.

LUW (Landbouw Universiteit Wageningen/bureau Blauw) (1991) *Een onderzoek naar de relatie tussen stofbelasting en stofhinder*, No. R 487, Wageningen, The Netherlands.

M&P (Melzer en Partners) (1991) *Inventarisatie en cumulatie van geluidbelastingen ten behoeve van de Integrale Milieuzonering IJmond*, No. PW.90.1.1., Dienst Milieu en Water Provincie Noord-Holland, Haarlem, The Netherlands.

OP&P (Oliemans Punter en Partners) (1993) *Leefsituatieonderzoek in de IJmond*, proj.code 4-035, Dienst Milieu en Water Provincie Noord-Holland, Haarlem, The Netherlands.

Provincie Noord-Holland (1981) *Provinciaal Milieubeleidsplan Noord-Holland 1982-1986*, Provincie Noord-Holland, Haarlem, The Netherlands.

Provincie Noord-Holland (1991) *Proefproject Integrale Milieu Zonering IJmond*, Project-programma, Provincie Noord-Holland, Haarlem, The Netherlands.

Provincie Noord-Holland (1992) M.M.M. van der Meij, afd. Milieu Technisch Onderzoek, Haarlem, The Netherlands.

Provincie Noord-Holland (1993) *Deelnotitie Fijn Stof*, Dienst Milieu en Water, Provincie Noord-Holland, Haarlem, The Netherlands.

SAVE (1991) *Integrale Milieuzonering IJmond, Externe Veiligheid*, Dienst Milieu en Water Provincie Noord-Holland, Haarlem, The Netherlands.

TNO (1991) Instituut voor Milieuwetenschappen TNO, *Integrale Milieuzonering IJmond, aspect Lucht, rapportage fase 1A*, (R91/188), TNO, Delft, The Netherlands.

TNO (1992) Instituut voor Milieuwetenschappen TNO, *Integrale Milieuzonering IJmond, aspect Lucht, rapportage fase 1B*, (R92/160), TNO, Delft, The Netherlands.

TNO (1992) Instituut voor Milieuwetenschappen TNO, *Integrale Milieuzonering IJmond, aspect Lucht, rapportage fase 1B*, (R92/160), VROM supplement, TNO, Delft, The Netherlands.

VROM [Dutch Ministry of Housing, Spatial Planning and Environment] (1978) *Noise Abatement Act*, The Hague.

VROM [Dutch Ministry of Housing, Spatial Planning and Environment] (1988) *Project 'Cumulatie van bronnen en integrale milieuzonering'*, The Hague.

VROM [Dutch Ministry of Housing, Spatial Planning and Environment] (1990) *Ministeriële handreiking voor een voorlopige systematiek voor de integrale milieuzonering*, Publ. reeks IMZ, No. 6., The Hague.

VROM [Dutch Ministry of Housing, Spatial Planning and Environment] (1991) *Basisdocument Stank*, The Hague.

VROM [Dutch Ministry of Housing, Spatial Planning and Environment] (1991) *Integrale milieuzonering*, rapportnr 17, The Hague.

VROM [Dutch Ministry of Housing, Spatial Planning and Environment] (1993) *Nota Stankbeleid*, The Hague.

Chapter 16

Integrated Environmental Zoning and American Inner-City Redevelopment: A Helpful Intersection of Land-use Planning and Environmental Law

M.A. Wolf [1]

16.1 A Hypothetical Case Study in Urban Distress

Metroville, a city with a population of 750,000, located in a metropolitan area with more than three million residents is, unfortunately, a typical American central city. In the older neighbourhoods comprising Metroville's inner city, much of the substantial industrial and warehousing building stock constructed in the early 20th century lies abandoned or severely under-used. Housing stock is in even worse shape, with hundreds of condemned and boarded-up tenement buildings, and tens of thousands of apartment units housing unemployed and underemployed low-income residents. Most of the large retailers have moved out to suburban malls, along with the great majority of professionals and white-collar workers. African-Americans, Hispanics, and Asians comprise more than fifty per cent of the city's population (the figure approaches ninety per cent in the inner city), and nearly one-third of the city's families live at or below the poverty line. The prevailing land-use planning pattern for the past few decades has been to segregate residential, commercial, and industrial uses, although some mixed-use parcels are present, either through pre-zoning non-conformities or as a result of small-scale zoning changes. The local government has decided that economic development – that is, providing jobs – is the key to revitalising Metroville's most distressed neighbourhoods. City officials are anxious to participate in federal programs that will assist Metroville in its redevelopment struggle. Some community activists warn, however, that the health and welfare of local residents cannot be sacrificed to this new development initiative.

16.2 IEZ and Urban Redevelopment

Over the past several decades, American governments at all levels – federal, state, and local – have experimented with a wide range of programs targeted to revitalising depressed inner-city neighbourhoods, such as those found in Metroville (Wolf 1987,

p. 77; Haar and Wolf 1989, pp. 925-964). A intriguing initiative was the Clinton administration's Empowerment Zones and Enterprise Communities (EZ/EC) plan, designed to create an environment for attracting significantly increased investment and employment to high-poverty regions (President's Community Enterprise Board 1994). Standing in the way of success for many potential EZ/ECs are severe disincentives to investment, including serious environmental hazards. Integrated Environmental Zoning (IEZ), devised to 'create [...] a framework for finding an integral balance between the environmental quality and the spatial-functional structure desired in the region' (VROM 1990, p. 1), presents a provocative method for measuring and accommodating environmental spillovers in an urban setting, particularly when industrial, commercial, and residential uses overlap.

16.3   Cities in Crisis: Diagnosis and Treatment

Many of America's central cities, like our hypothetical Metroville, are dying. The warning signs are hard to miss: decaying infrastructure, declining tax bases, middle-class flight to the suburbs, polluted rivers and harbours, poisoned air, skyrocketing crime rates and insurance premiums, private sector disinvestment, outdated land-use plans, and government neglect. In several cases, those who are unable or unwilling to flee the congestion, crime, and pollution are poor and working class African-Americans, Hispanics, and Asians, a demographic correlation that has been noted of late by those active in the environmental justice movement (Commission for Racial Justice 1987).

There are two basic solutions to this severe crisis: allowing the disease to take its course or reinvigorating the patient. Advocates of the first course suggest that attempting to revive cities now with public or private dollars is fruitless. If instead we allow conditions to continue to deteriorate, real estate values will fall to levels that are so low that investors and speculators will begin to assemble parcels for large-scale development. At that point central cities would probably no longer serve as residential centres, a trend that has already been set by some cities in the West and Southwest. The most apparent and devastating impact of this strategy would be the dislocation of hundreds of thousands of inner-city residents and the destruction of entire communities, their political power bases, churches, and social centres.

Policymakers who instead champion the active reinvigoration of America's inner cities generally endorse one of two vehicles. Those who gain their inspiration from the grand schemes of the New Deal of the 1930s and the Great Society of the 1960s look to government spending to cure urban ills (Rohatyn 1994). However, given budgetary constraints, the relatively weak political influence of the urban poor, and the overwhelming public resistance to significant tax increases, there is little likelihood that an ambitious federal spending program will emerge anytime soon. On the state and local levels – where budget restrictions are even tighter – there is an even smaller probability.

The second group of advocates for inner-city rebirth see private investment as the only effective vehicle. Businesses and investors should be encouraged to provide jobs, start-up capital, and loan support for residents and entrepreneurs in the nation's most distressed communities. The public-private joint venture, not public works, is the new paradigm (Haar and Wolf 1984, pp. 925-929).

In the 1980s the program that best embodied this concept was the enterprise zone (EZ). EZs, imported into the U.S. from the United Kingdom by the conservative Heritage Foundation, involved the use of tax, financing, and regulatory incentives to attract increased employment and investment to the nation's most distressed regions (Butler 1981; Wolf 1990, pp. 125-126). By the beginning of the 1990s more than forty states had enacted some kind of EZ legislation, and in more than thirty states EZs were in operation, offering a wide variety of incentive packages to employers and investors (Wolf 1989).

Although EZs had Republican ties from the beginning, it was not until the election of a Democratic president – Bill Clinton – that the federal government passed an EZ program. In August 1993, Congress authorised the Department of Housing and Urban Development to designate six urban Empowerment Zones and sixty-five urban Enterprise Communities during the subsequent two years. Empowerment Zones, urban neighbourhoods with poverty rates exceeding twenty per cent, would be eligible to receive a significant injection of federal funds ($100 million per zone) and to offer income tax credits to employers in the zone. As the second tier of the program, Enterprise Communities will receive $3 million each and other tax incentives as well. In addition, the Clinton administration has promised technical support, lending and regulatory assistance, and priorities for a wide range of federal programs in fields such as transportation, education, health care, social services, and criminal justice. This aid is available not only in EZ/ECs, but in many cases also in program-eligible areas that compete in the nomination process but are not selected (The President's Community Enterprise Board 1994).

The strong governmental presence in EZ/ECs is based on the realisation that there are some severe disincentives to investment in decaying inner cities that cannot realistically be overcome simply by letting the private sector 'run free'. Chief among the barriers that cannot be overcome simply by reducing taxes or erasing regulations are the scarcity of a trained, educated work force; widespread public safety concerns; crumbling infrastructure; prohibitively high insurance rates; environmental abuse; and outdated land-use plans. It is to the last two disincentives, which are of course most relevant to a discussion of IEZ in the inner city, that we now turn our attention.

16.4   Environmental Barriers to Inner-City Redevelopment: Accommodating Public-nuisance-like Spillovers

Many of America's largest cities have high pollution profiles, a reality that makes redevelopment initiatives problematic if not unrealistic. For example, according to the figures released by the Environmental Protection Agency (EPA), more than 50 million Americans live in areas that do not meet federal air quality standards (U.S. Environmental Protection Agency 1993, p. 1-14). Nor does the situation get cleaner when we shift media, as many American cities have been struggling with the dirty legacy of rapid residential, commercial, and industrial development from decades past: polluted harbours and estuaries, inadequate waste treatment, uncontrolled and untreated runoff, landfills at or near capacity, and abandoned industrial sites with toxics in or below the soil.

The predominant pre-regulatory legal mechanism for contending with severe environmental harm was public nuisance, a legal action derived from the prosecution of individuals who conducted activities that posed harm to the 'common' or to the 'community at large' (Keeton 1986, pp. 617-618). While public nuisance actions remain available in many instances, today provisions for fines, abatement, and imprisonment for those committing similar offences against the public good are included in the modern environmental statutes and regulations that began to appear in the 1970s.

To employers and investors interested in locating in many of America's inner cities, the pollution-control mandates and standards included in environmental statutes and regulations can dramatically affect the bottom line. New significant sources of air pollution located in non-attainment areas, for example, are required to use technology that will yield the lowest achievable emissions rate (LAER) (Percival 1992, p. 802). A new private facility that discharges pollutants into bodies of waters deemed 'toxic hot spots' under the Clean Water Act may well need to meet more demanding pre-treatment standards than are required of competitors that release into cleaner rivers and streams (Percival 1992, pp. 939-943). Similarly, industrial and commercial newcomers are often asked to pay costly hook-up fees to tie into state-of-the-art waste treatment facilities mandated by federal law. Even if these significant disincentives can be overcome, the presence of hazardous materials would necessitate the expenditure of considerable funds to clean up many urban industrial and commercial sites before redevelopment could begin.

If federal officials are serious about their offer to play an active partnership role in inner-city revitalisation, they must be willing to respond in meaningful and effective ways to these environmental externalities. There are two basic strategies: (1) providing low-interest loans or subsidies for the development and implementation of state-of-the-art pollution control devices, or (2) allowing waivers for employers, developers, and investors who risk capital in truly distressed neighbourhoods. There is a down side to each of these options, however.

Because the first choice (grants and loans) is not 'revenue-neutral', legislators will have to expend political capital and make some hard choices (for example, downsizing or eliminating programs competing for the same dollars, increasing taxes) before the federal or state monies are available. Given current budget limitations and the public's reaction to revenue increases, there is little likelihood of significant movement in this direction. These financial constraints make the second option more desirable, at least on first glance.

The second option (allowing waivers of environmental requirements), though free of budget and revenue constraints, is problematic for a different reason. Owing to a significant grass-roots movement, there has been a growing awareness that members of minority groups, indigenous peoples, and the urban and rural poor are often forced to carry too heavy a burden as a result of environmental decision-making by the public and private sectors. President Clinton responded to calls for 'environmental justice' by issuing a wide-ranging executive order that reads in part; 'To the greatest extent practicable and permitted by law, [...] each Federal agency shall make achieving environmental justice part of its mission by identifying and addressing, as appropriate, disproportionately high and adverse human health or environmental effects of its programs, policies, and activities on minority populations and low-income populations in the United States [...]' (Clinton 1994, p. 7629).

Because of the high concentration of African-Americans, Hispanics, and Asians in the nation's most distressed inner-city neighbourhoods, federal agencies would be most critical of any environmental waiver in Empowerment Zones and Enterprise Communities that would even appear to expose residents to increased risk.

The IEZ approach can help to offset some of the negative repercussions of the second option. Mapping the area targeted for economic development and measuring the ways in which noise, odours, and poisons in the air dissipate over distances can in some cases assist local planners in creating a revitalised neighbourhood in which residents and their commercial and industrial jobs are not separated by great distances. In other instances, the data generated by IEZ will support decisions by local governments to reject development proposals that would put special burdens on the urban poor. Such assistance would be a welcome addition to local land-use decision-making. For, as the following section indicates, although American analogues to IEZ have appeared over the past few decades, planning and zoning continue to pose additional barriers to redevelopment initiatives.

16.5 Planning Barriers: Accommodating Private-nuisance-like Spillovers

Most of the negative externalities involved in siting commercial and industrial uses in close proximity to residences do not approach the 'public harm' level we equate with public nuisance and its modern legacy – comprehensive environmental statutes and regulation. Instead, the inconveniences occasioned by noise, odours, traffic, and vibrations are more readily identified with the civil case brought against one who unreasonably interferes with the use and enjoyment of another's real property – that is, private nuisance (Keeton 1986, pp. 619-620). In many ways, we can view local zoning and planning as the modern regulatory legacy of private nuisance, as planners sought to segregate, rather than outlaw, incompatible uses into separate zones. Indeed, in judicial decisions upholding the validity of height, area, and use zoning ('Euclidean zoning'), judges often referred to principles of private nuisance law for guidance and legitimacy (Haar and Wolf 1984, p. 184; Mandelker 1988, p. 56).

For decades, this exclusion and separation made sense in the new American suburb, the 'bedroom community', in which the primary goal of planners was to protect the sanctity of the single-family, detached dwelling neighbourhood. Cities and older suburbs posed special problems, however, and planners were forced to contend with significant numbers of non-conforming uses, the growing popularity of high-rise apartment buildings, and the reuse of warehouse and industrial properties for commercial and residential purposes. It soon became clear that without modifications

Euclidean zoning would not be up to the task of accommodating competing demands for limited urban space.

Those modifications took several shapes. First, planners attacked the idea of cumulative zoning. While Euclidean zoning allowed less intensive uses in more intensive use zones (for example, homes in areas zoned for business or industry), some planners advocated exclusive industrial and commercial zones, to preserve the integrity and logic of the comprehensive plan, to enhance the property tax base, and to protect more intensive users from ultra-sensitive neighbours. Although some courts were troubled by non-cumulative zoning (particularly when shopping centres were banned from industrial zones), in many cases judges have allowed planners to exclude residences from the zones envisioned for the most intensive uses (Mandelker 1988, pp. 163-165).

One major city – Chicago – has experimented with Planned Manufacturing Districts (PMDs), in an effort to halt conversion of industrial structures into businesses and residences. The success or failure of this program is the subject of heated debate in Chicago. Advocates see the districts as an effective part of the city's strategy to maintain its industrial base, while opponents complain that the city can not dictate uses that the market does not drive (Strahler 1993).

A second modification of Euclidean zoning's height, area, and use classification scheme has been the implementation of primitive performance controls. Targeted especially at industrial noises, odours, and vibrations, these non-quantitative standards have also passed judicial muster, despite allegations that phrases such as 'offensive' and 'excessive' are too vague and ambiguous. However, it is just this 'subjective' quality of primitive controls that has led to the demand for expert-based standards (Haar and Wolf 1989, pp. 276-279; Mandelker 1988, p. 162; Acker 1991).

During the 1970s, the federal government encouraged the development of objective, expert-based standards to measure and control excessive noise, as part of the Noise Control Act of 1972 and the Quiet Communities Act of 1978. In accordance with this legislation, and with the technical and financial assistance of the federal government, several states and localities developed noise ordinances. This federal-state-local joint venture fell out of political favour in the early 1980s and state and local funds decreased dramatically as federal funds dried up (Shapiro 1992, pp. 1-3).

A third departure from traditional Euclidean zoning involves various devices designed to bring flexibility to an otherwise rigid scheme of classification and segregation. Transferable development rights, for example, enable local governments to protect environmentally sensitive and historically significant properties by shifting development to more appropriate sites elsewhere, while maintaining the same overall density. A growing number of localities are granting zoning changes conditioned upon the landowner's promises to provide public amenities, to shield adjoining neighbourhoods from offensive sights and sounds, to restrict potentially disturbing activities that otherwise might be conducted on the site, and even to make cash payments to offset increased governmental costs attributable to the development. Some of the representations made by developers seeking this 'conditional zoning' can be recorded in the public registry as legally enforceable restrictive covenants, thereby enabling the local government or adjoining neighbours to ensure that promises are kept

(Haar and Wolf 1989, pp. 256-276, 283-288; Mandelker 1988, pp. 262-267, 464-465).

An IEZ initiative in Metroville could build on and enhance these 'post-Euclidean' planning and zoning devices. First, Geographical Information Systems (GIS) mapping and objective data on dispersal of environmental disturbances could provide planners with harder data than is currently available (Mank 1992, pp. 777-791). As a result, exclusive industrial and commercial zones would be no larger than required for the anticipated development; this in turn increases the chances that inner-city residents will be able to live close to their jobs. Second, IEZ could provide much-needed usable standards that are missing from primitive performance zoning schemes. Third, the usable standards generated from the IEZ process could be documented and recorded for the protection of nearby residents and the locality. In this way, IEZ would fit in well with efforts to make zoning and planning more flexible and more responsive to the special needs of today's urban neighbourhoods.

16.6 IEZ as a Helpful Intersection between Land-Use Planning and Environmental Controls

Generally in the United States, land-use planning remains a local and state responsibility, while the federal government has taken the lead in crafting and implementing environmental statutes and regulations. There are, of course, exceptions, such as the Federal Land Policy and Management Act of 1976, mandating land-use planning for federal lands, and several state and a few local environmental protection acts, requiring the preparation of environmental analyses before development may proceed (Haar and Wolf 1989, pp. 727-736; Percival 1992, pp. 1124-1126). Still, Congress and federal agencies remain the primary environmental policy makers, while local (and to a lesser extent, state) officials have the most important say when it comes to planning and zoning.

The federal noise control program, when funding was in place, was a good example of co-operation among the various governmental branches. Other provisions of federal environmental acts regarding topics such as wetlands, floodplains, air pollution, and endangered species often have a direct impact on local planning decisions. For the most part, local planners, particularly in central cities with severe budgetary restraints, often lack the kind of technical assistance required for developing, implementing, monitoring, and enforcing their own environmental controls. Indeed, too many local planners are already overburdened with performing their 'traditional' functions.

Unfortunately, these shortcomings have not stopped some suburban and ex-urban localities from asserting environmental rationales for their planning and zoning decisions. For example, in some of the leading cases in which local governments were accused of erecting one- or two-acre lot minima in order to exclude lower-income residents from the inner city, local officials raised ecological factors – aquifer protection, sewage treatment, and flooding and drainage – without any apparent data to back up their claims of impending harm (Haar and Wolf 1989, pp. 390, 405-406). These suburban exclusionary zoning cases thus raise some of the same environmental justice concerns that we normally associate with inner cities and poor rural communities.

IEZ, if it is carefully crafted, administered, and monitored, presents local governments with the opportunity to include selected environmental factors in planning decisions. A word of caution is in order, however. IEZ experimentation in America should first be confined to those private-nuisance-like activities that gave rise to the segregation of incompatible uses inherent in Euclidean zoning. That is, by measuring and mapping the dispersal of non-dangerous noise, odours, and vibrations, planners can provide the kind of fine tuning needed to make zoning more responsive to inner city redevelopment initiatives designed to bring jobs and investment to urban neighbourhoods.

In the specific context of EZ/ECs, as in state enterprise zones, IEZ could be especially helpful, for many of the most important incentives are available to businesses located in the zone who hire workers who live and work in the zone. In many cases, the data derived through IEZ could enable zone businesses to assure the community that they could conduct their commercial and industrial activities and still be compatible neighbours. In other instances, the data could indicate that the specific commercial or industrial activity envisioned for a parcel would be too intrusive, leaving the landowner with a choice between acquiring additional property as a buffer (depending on state law, the local or state government might be able to use its power of eminent domain to assist this economic development effort), investing in equipment to reduce the spillover, or operating (or selling to someone else who would operate) a less intensive use.

Only after IEZ has a track record in this private-nuisance-like context should we begin to explore expanding application in the area of public-nuisance-like activities that may pose significant harm to the human and non-human environment. At this point, we cannot be sure that either the technology or the local government expertise is up to these tasks. Moreover, particularly in the inner-city setting in the nation's real Metrovilles, the environmental justice implications of the premature employment of a new environmental planning methodology are serious enough to warrant extreme caution.

Note

1   Michael Allan Wolf is Visiting Professor of Law at the American University and Professor of Law and History, University of Richmond, USA.

References

Acker, F.W. (1991) Performance Zoning, *Notre Dame Law Review*, Vol. 67, pp. 363-401.
Butler, S.M. (1981) *Enterprise Zones: Greenlining the inner city*, Universe Books, New York, NY.

Clinton, W.J. (1994) Federal Actions to Address Environmental Justice in Minority Populations and Low-Income Populations, Executive Order 12898 of February 11, 1994, *Federal Register*, Vol. 59, pp. 7629-7633.

Commission for Racial Justice United Church of Christ (1987) Toxic Wastes and Race in the United States: A National Report on the Racial and Socio-economic Characteristics of Communities with Hazardous Waste Sites.

Haar, C.M., and M.A. Wolf (1989) *Land-Use Planning: A Casebook on the Use, Misuse and Re-Use of Urban Land*, Little, Brown and Company, Boston, MA.

Keeton, W.P. (1984) *Prosser and Keeton on Torts*, West Publishing Co., St. Paul, Minnesota.

Mandelker, D.R. (1988) *Land Use Law*, The Michie Company, Charlottes-ville, Virginia.

Mank, B.C. (1992) Preventing Bhopal: 'Dead Zones' and Toxic Death Risk Index Taxes, *Ohio State Law Journal*, Vol. 53, pp. 761-804.

Percival, R.V., et al. (1992) *Environmental Regulation: Law, Science, and Policy*, Little, Brown and Company, Boston.

The President's Community Enterprise Board (1994) *Building Communities: Together, Empowerment Zones & Enterprise Communities Application Guide*, U.S. Department of Housing and Urban Development, Washington, DC.

Rohatyn, F. (1994) Public Works for Urban Ills, *Newsday*, March 24, p. A44.

Shapiro, S.A. (1992) Lessons from a Public Policy Failure: EPA and Noise Abatement, *Ecology Law Quarterly*, Vol. 19, pp. 1-61.

Strahler, S.R. (1993) Protection Racket: PMDs Revisited, *Crain's Chicago Business*, November 8, p. 17.

U.S. Environmental Protection Agency (1993) *National Air Quality and Emissions Trends Report, 1992*, Research Triangle Park, North Carolina.

VROM [Ministry of Housing, Physical Planning and Environment] (1990) *Ministerial Manual for a Provisional System of Integral Environmental Zoning*, The Hague.

Wolf, M.A. (1987) Potential Legal Pitfalls Facing State and Local Enterprise Zones, *Urban Law and Policy*, Vol. 8 (1986-1987), pp. 77-130.

Wolf, M.A. (1989) An 'Essay in Re-Plan': American Enterprise Zones in Practice, *Urban Lawyer*, Vol. 21, pp. 29-53.

Wolf, M.A. (1990) Enterprise Zones: A Decade of Diversity, in R.D. Bingham et al. (ed.), *Financing Economic Development*, Sage Publications, Newbury Park, CA, pp. 123-141.

Chapter 17

(Integrated) Environmental Zoning: A Comparative Study of 12 Countries

J. Pearce[1]

17.1 Introduction

During 1992 a detailed comparative survey was carried out in 12 different countries worldwide (Lloyd's Register 1993). This survey was carried out as part of a program of developing policy instruments in The Netherlands known as *environmental zoning* and *integrated environmental zoning*. The study was commissioned by the Netherlands Ministry of Housing, Planning and Environment, and carried out by a team involving Ministry personnel, Lloyd's Register and two independent consultants. The purposes of the study were, firstly, to check the extent to which similar instruments (or components of them) were being developed in parallel in other countries or whether in fact the Netherlands was going out on a policy limb, and secondly to look at any problem areas encountered, so that lessons can be drawn which would help in developing Dutch policy. The environment is becoming an increasingly international matter as a result of the realisation that many issues facing us are either intrinsically international or have an international dimension, and in Europe as a result of the increasing extent to which environmental policy is being made at the level of the European Union. This means that European countries are increasingly trying to ensure that national initiatives are consistent with developments in partner countries. Even instruments with a relatively local scope, such as environmental zoning, are not immune from this concern.

Environmental zoning is an approach to tackling problems of local pollution and nuisance around industrial estates caused by the close juxtaposition of industry (or in principle other concentrations of nuisance) and housing. A closed curve is drawn around the estate, as illustrated by Figure 17.1. The area enclosed by this curve is referred to as the environmental zone. Outside of the zone the environmental loading must not exceed a set value, or standard. Inside the zone, restrictions are placed on the building of housing and other environmentally-sensitive functions. In the case of existing industrial zones, the line is the product of a negotiation process, but normally corresponds to a given estimated environmental iso-quality contour applying either now or at some time in the future allowing for realisation of expansion projects.

The quantified nature of both the determination of the configuration of the zone and the environmental quality generally around the estate is crucial, as it makes the zoning process space-efficient. Environmental zoning therefore recognises that it is not always possible at present to meet the increasingly strict standards that society expects of environmental quality at the factory fence, and that a buffer zone is sometimes required to separate environmentally intrusive activities from environmentally sensitive activities.

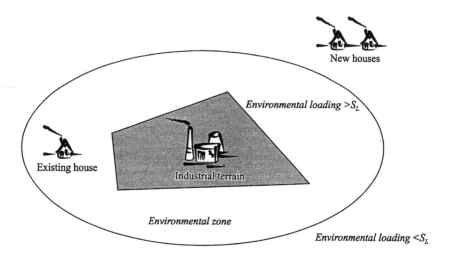

Figure 17.1   Environmental zoning around an estate

Environmental zoning is an instrument involving not only environmental policy, but also physical planning, since it places restrictions on planners in assigning land use. It therefore requires co-ordination between these two policy functions. It would of course be at its most effective for dealing with a green field situation. Unfortunately for policy-makers the industrial topography in industrialised countries is largely defined already, and this means that schemes for environmental zoning have to contend predominantly with existing situations, and in particular houses lying within the environmental zone which is drawn, and therefore subjected to environmental loadings higher than those judged acceptable for new housing. In The Netherlands therefore, zoning systems have invariably been accompanied by abatement schemes for ensuring that existing housing, though not meeting the standards for new buildings, are nevertheless not subjected to excessive nuisance. This also ensures that the trade-off between setting land-use restrictions and taking measures to reduce emissions are not weighted too heavily in favour of the former.

Environmental zoning has been used in a number of contexts in The Netherlands, the most notable being in the Noise Abatement Act of 1979. As part of the measures to control industrial noise nuisance, noise zones, i.e. environmental zones in the above sense based on noise loadings, were set up around some 750 industrial estates, and noise abatement schemes were set up around 500 estates (VROM 1989).

Noise is of course not the only environmental impact which industrial estates have on nearby communities. In The Netherlands, odour, for example, is a greater nuisance if numbers of complaints occurring is taken as the criterion. Given the groundswell in favour of taking an integrated approach to environmental problems, *integral* environmental zoning (IEZ), in which the loading criteria are based on all the relevant environmental components, and not just a single component such as noise, is the logical development.

An IEZ project was therefore started at the Ministry in 1989. Environmental components included were noise, odour, toxic and carcinogenic air pollutants and major hazard. Work carried out to date includes:

a inventory of zoneable sites;
b ongoing work on the development of methods for aggregating different types of environmental loading together to obtain a composite environmental quality indicator;
c pending (b), establishment of a provisional system of IEZ;
d costing of associated abatement for existing situations;
e establishment of ten pilot IEZ projects at a number of sites around The Netherlands.

17.2 The Environmental Zoning Survey

The environmental zoning survey was carried out in the course of 1992. Restricted resources meant that only a limited number of countries could be included and these were carefully targeted, based on a number of criteria, including prior indications about the relevance of particular countries, and a necessary Eurocentricity. The countries selected were:

- Australia
- France
- Germany
- Hong Kong
- Italy
- Japan
- Singapore

- Spain
- Sweden
- United Kingdom
- United States (California, Washington, NY State, Ohio)
- (The Netherlands)

After preliminary research, each country was visited, and meetings were held where possible and relevant with government officials involved both in environmental protection and in planning both at central and local levels, with specialists in environmental law and with environmental specialists working with large corporations. The ground covered by the interviews included:

Conflicts

- To what extent is (integral) environmental zoning, or elements of it, to be found in the different countries?
- Juxtaposition of industrial estates with housing, how important is this issue, how has it developed historically, and what instruments are used to deal with it?

Instruments

- standards;
- Environmental Impact Assessment.

Management

- the way responsibilities are assigned between different government departments;
- approach to environmental management for industry;
- relationships between environmental management and physical planning;
- elements of the land-use planning system.

17.3   The Results of the Survey

The environmental zoning survey illustrated that situations in which residential areas are in close proximity to polluting industry are a universal phenomenon in industrialised countries. Instances of this were encountered in all the countries visited.

In countries as diverse as Japan, Hong Kong and The Netherlands, the great majority of complaints about environmental matters relate to noise and odours. These are local nuisance problems, usually (in the case of odour) or often (in the case of noise) attributable to nearby industry.

Although such situations sometimes arise as a direct result of the establishment of a new factory at an insufficient distance from existing housing, the most common circumstances are either:

- that the factory(ies) date from an era when the proximity of the workforce to the workplace was a paramount consideration, and when a clean environment took a very secondary place, or
- that the distance between industry and housing, although once adequate, has been eroded by encroaching housing.

Amongst environmental regulators, planners and industrialists there was a widespread and explicit recognition that the issue of the close juxtaposition of houses to locally polluting industry is an important environmental issue, particularly at the local level. There was also a general (but not universal) recognition that source-oriented mitigation measures are not in themselves always sufficient to deal with these problems. Other instruments, particularly physical planning, are being developed to prevent these situations from arising.

No cases of a fully IEZ-like approach (that is, a buffer zone between industry and houses, with the configuration of the buffer zone depending on actual environmental quality, physical planning restrictions inside the zone, and the integration of different pollution types) were found anywhere in the world. Schemes containing some of these elements, however, are being used in many of the countries.

Some of the ways of trying to deal with the juxtaposition problem are considered further below.

Environmental Quality Standards

Environmental quality standards are obviously a direct way of trying to protect environmental quality in residential and other areas. Quality standards exist in some form or other in all of the countries considered, although there are problems in standard-setting, particularly in regard to odour and hazard. These standards alone cannot achieve what is achieved by environmental zoning, however.

First of all, environmental quality standards, by their nature, tend to take a non-integral approach to environmental quality. That is, they consider each environmental quality component separately. Although there is no reason why a composite standard, involving a number of different environmental components, should not be developed and applied, no examples were found of such integrated standard-setting. Secondly, environmental standards are not normally binding on planners, so they are not effective in preventing the encroachment of housing towards industry. Quality standards are not able to come to terms with the dilemma that a satisfactory environmental quality is not always feasible in technico-economic terms at the factory fence.

Environmental Impact Assessment (EIA)

There are a number of parallels between EIA and IEZ, although the scope of EIA is wider than IEZ, in that it looks at all impacts and not just selected quality parameters in residential areas. It can be an instrument for ensuring that new developments, including industrial developments, do not result in an unacceptable deterioration in environmental quality in living areas. It gives planners access to detailed information about the environmental consequences of new developments, and allows an informed trade-off to be made between environmental and physical planning.

In some countries, the EIA process feeds back into the locational decision making, so that the environmental and physical planning trade-offs are made more explicit. In order for this to be possible, the EIA must be carried out for a number of alternative sites, or partly in a non-site-specific way. An example of this is Sweden, where an industry seeking a building licence for a new site is obliged to submit details of a number of alternative sites. Alternatively, the process may work in an iterative way, with the realisation that some environmental impacts are unacceptable leading the applicants to reconsider alternative locations or consider further abatement measures for emission sources.

EIA can ensure that the necessary separation of industry and housing occurs when new industries are being established, particularly where there is a comprehensive system of environmental quality standards. Since it only looks at impacts *outwards* and not impacts *inwards*, it only has a once-and-for-all value, and does not protect against subsequent encroachment by housing towards the factory.

Ad Hoc and Single-Component Buffer Zones

Although buffer zones calculated within the systematic and integrated framework of an IEZ-like system hardly exist yet, a variety of more ad hoc or single-component schemes are to be found which seek to ensure a separation between industry and housing. These are discussed below:

Buffering through Land-Use Planning

Detailed land-use planning at a local level is carried out almost universally. The drawing-up of a land-use plan permits a separation of environmentally intrusive and environmentally sensitive activities to be achieved by means of the interposition of an environmentally neutral activity or a green belt between them. Office accommodation might be regarded as an example of such an environmentally neutral function. Examples of the effective use of land-use planning to achieve separation by these means were seen in many countries, and are clearly widely applied. Use may also be made of natural or infrastructural features to reduce the approach of housing and industry to each other.

This practice is not limited to public planning. Managers of large industrial estates, working in concert with environmental regulators, are able to apply a grading scheme in allocating plots to industrial tenants such that the more polluting industries are situated in the inner areas, and therefore separated from housing and other sensitive functions by the less polluting industry.

Land-use planning alone is unfortunately a rather limited way of achieving the necessary separation. Land-use categories are often not sufficiently discriminating, and the environmental data required to make it a more efficient system is usually lacking. Moreover, planners are subject to too many constraints and pressures to ensure that this is consistently done.

Generic Buffer Zones

In some countries, 'generic' buffer zones have been established. That is, environmental zones have been established with a radius which is determined for categories of sources, usually those belonging to a particular industry type. These generic buffer zones make an assumption about the mean distance within which the environment is likely to be rendered unacceptable by the various types of emission and nuisance which such an installation causes. This distance is based on experience with similar establishments in other situations.

Systems of this nature were found in the state of Victoria in Australia, in Sweden, in the German state of Nordrhein-Westphalia, in Hong Kong, and in Singapore. There may be a provision to decrease or increase these distances in specific cases. There is a great range in the bindingness of these buffer zones, and the importance allocated to them. In Sweden and in Hong Kong, for example, they are guidelines only, and not necessarily enforced, whereas in Victoria and in Singapore the buffer zones are effectively written into the land-use plans.

In Hong Kong the buffer distances are a function not only of the type of industry, but also of the nature of the environmentally sensitive object, for example housing, old people's home, etc.

Buffer zones set in this manner are a compromise between environmental efficiency and convenience of application. Because the distances are generic, there will be cases where land is needlessly subjected to restrictions in its use, and other cases where the environmental quality is not satisfactory even beyond the buffer zone. To be set against these disadvantages is the advantage that the buffer zones are simple to assign, and that the extensive collection of data, calculation, and negotiation is not required.

Single-Component Buffer Zones

A number of environmental zoning schemes have been developed which aim to bring a separation between housing and industry in order to manage a single environmental component only. These separations are analytically determined on the basis of agreed criteria, and most often relate to major hazard.

Examples are the 'consultation zones' in the UK and in Hong Kong, 'areas of concerted policy' in France, the zones established around hazardous installations in New York state and around LPG installations in Washington state and in The Netherlands, and land-use restrictions applied around hazardous installations in Victoria and New South Wales.

The principle of environmental zoning appears to have proved more acceptable for major hazard than for other environmental components. Major hazard is of course a different type of phenomenon from those such as odour, air pollution and noise. Major hazard is a potential (admittedly catastrophic) environmental intrusion rather than an actual one, and in most cases will never be realised. If it does occur, then the effects both in human safety and political terms may be large. This may be the reason why decision-makers are willing to take physical planning measures as well as technical abatement measures to mitigate the situation.

Buffer Zones Created by Industry Itself

Polluting industry is also incommoded by the existence of housing or other sensitive functions in its close neighbourhood. Complaints and even lawsuits (particularly in the US) are more likely, and the observance of quality standards directed towards residential areas are more difficult to meet. Industry therefore often strives to create its own buffer zones around polluting manufacturing plants, with or without the encouragement of the regulatory authorities.

This can be done, particularly in the case of larger factories, by situating the most intrusive installations as far as possible from the perimeter or the position where houses are to be found. It is also done by means of the purchase of surrounding land or houses subjected to high environmental loading or, in the case of new industry, by the pre-emptive purchase of larger plots of land than that required. This is particularly practicable where land availability is good, for example in the United States and Australia, but is also common in Europe.

Purchase of land surplus to immediate requirements also provides industry with additional space for future growth if needed.

Consultations between Planning and Environmental Functions

Although planners are not generally required to apply environmental standards directly, they are sometimes required to consult with environmental authorities before approving certain planning applications.

Relocating Industry or Housing

Environmental zoning is easiest to apply and most appropriate where a separation already exists between industry and housing, as a means of protecting against the future industrial or residential erosion of this separation. In practice, policy-makers often have to deal with existing situations where encroachment has already occurred or where such a separation never existed. A very widely practised means of dealing with existing problem situations which cannot be tackled purely by means of abatement measures at the source is to relocate the industry or relocate the houses.

Relocation of industry in particular is a widespread practice, especially for dealing with smaller but intrusive industries situated in cities or other densely populated areas. The mechanisms by which this occurs are various. Pressure from the regulatory authorities is an important factor, and the need for renewal of licences may provide an opportunity for the exercise of such pressure. Complaints from the local community will increase this pressure, and since by definition in such situations the space for expansion may be limited or non-existent, the establishment may be obliged to relocate because of expansionary pressures.

The main obstacle to moving industry (or housing) is the costs involved. In Japan, in particular, financial incentives are often provided to encourage industry to move. In both Tokyo and Yokohama, considerable subsidies were available on the interest on loans raised for relocation purposes. In Hong Kong the value of residential land commands such a premium over that for industrial use that the full costs of relocation can be met by the increase in value of the land released. A similar phenomenon can be observed in Singapore.

The mechanism by which the relocation of housing occurs is frequently the purchase of the housing by industry, as seen above. This is often preceded, in the case of houses near heavy industry, by a natural dynamic in which a deteriorating environment and changes in social values lead to a blighting and decline in value of nearby houses.

17.4   General Prospects for IEZ Systems in the Future

Do IEZ-like systems offer the key to dealing with problems of local nuisance and pollution caused by industry? The survey highlighted various obstacles to the introduction of IEZ as a practical policy tool. These can be considered under the following categories:

Legal

In most countries there appears to be no legal impediment to the introduction of an IEZ-like instrument. The notable exception to this is the USA, where at least in certain states the law and cultural factors are hostile to placing restrictions on the freedom to construct houses on owned land. Removing this freedom could give rise to claims for compensation under the Fifth Amendment or other litigation.

Economic

The most controversial cost category is that related to the reduction of land values within zones, as a result of restrictions on possible land use (land blighting). This can be particularly problematic where the zones are in urban areas where the value of development land is high, or in areas where the zones need to be large. Environmental zones which freeze the development of large expanses of prime urban land will be economically and therefore politically unacceptable. The key to resolving such problems seems to lie in ensuring the correct balance between source abatement measures and the restriction of land use. Industrial firms may need to undergo a pollution abatement program before environmental zones are established around them rather than afterwards, to ensure that these zones do not need to be too large.

Physical

The argument encountered particularly in Japan, but also to some extent in Hong Kong, is that where space is in short supply, the creation of buffer zones and the introduction of physical planning constraints as implied by environmental zoning is precluded by sheer lack of room. The further constraints which would be put on the physical planning by an environmental zoning system are simply too great, and would involve foreclosing too many options.

Technical

A system of environmental zoning with the elements (a) to (e) described in section one is data-intensive and study-intensive. Not only are extensive data required on air emissions, noise emissions and odours, but an elaborate edifice of agreed air dispersion and noise transmission models together with standardised olfactometric techniques, etc. are required.

Although extensive emission inventories are becoming increasingly common amongst the countries visited, these usually do not cover either noise or odours. Inevitably considerable study will be involved in the process of setting up such zoning systems. In countries such as Hong Kong, where licensing only covers some of the largest industrial installations, the database is presently also lacking for air pollution.

Several countries are trying to bypass this problem, while moving towards a system of buffer zones by making use of 'generic' buffer distances, that is buffer distances which depend only on the type of polluting industry and possibly the type of sensitive function involved. Such an approach is to be found in Germany (Nordrhein-Westphalia), in the states of Victoria and New South Wales in Australia, in Singapore and in Hong Kong. To compensate for the loss of efficiency, in particular that the zone will be made unnecessarily large, such schemes are often accompanied by a provision that a buffer distance can be reduced if special abatement measures are taken, or if it can be shown that the deterioration in environmental quality caused does not warrant such a large buffer zone.

Institutional

The implementation of a policy for environmental zoning, and its management, call for collaboration between planners and those responsible for environmental management. These two activities are often organisationally relatively remote from one another, at least at the level of central government. Even where they are part of a single ministerial portfolio (e.g. UK, Netherlands), they may be quite separate branches within the ministry. There is moreover an implicit conflict of objectives involved, with environmental planners seeking to ensure satisfactory environmental quality in residential areas, if necessary by means of physical separation, and the spatial planners seeking to minimise any further constraints on land use. Particularly when the availability of space for housebuilding is limited, planners may not welcome policy initiatives which further reduce their options. This applies particularly where the application of the given environmental quality standards implies a freeze on the development of large areas of land.

Another institutional issue which arises is that of central versus local interest. Environmental zoning is by its nature a local issue, and individual situations where it would be applicable vary widely from case to case, as do the economic, environmental and socio-infrastructural priorities applying. Detailed land-use planning is a local affair. On the other hand local government may not have either the legal powers or the political weight required to introduce such an instrument at the local level.

If initiatives are taken at the central government level to introduce the policy instrument, the issue arises as to whether these should be prescriptive or enabling, and if the former whether they should establish a uniform system, or provide for some flexibility to allow for local circumstances as referred to above. Attempts to establish a uniform system may lead to some unacceptable local situations.

Cultural

Cultural factors can be important in determining whether environmental zoning schemes are likely to be adopted in various countries, although it is of course sometimes difficult to disentangle cultural from legal, socio-economic and other considerations. The following are examples of such cultural factors which may operate.

In the US, cultural traditions are not sympathetic to zoning systems established by the public authorities. Regulations which restrict the use to which private land can be put, particularly where it prohibits the construction of houses, are regarded as infringing on private property rights. Where buffer zones have been established, this is often by initiative of the industries themselves rather than by local government.

The highly technocratic approach of IEZ (use of quantified standards, relatively deterministic calculation or configuration of zone) also runs counter to the tendency in some countries to take a more pragmatic approach. It may additionally be seen as diminishing the discretionary element of decision-making by locally elected officials. Since this discretionary capacity provides them with a basis for their power and authority, anything seen as assailing it may be opposed.

17.5 Conclusions

Whether IEZ catches on as an instrument for dealing with local environmental problems depends on the weight attaching to the obstacles, and how they weigh up against the undoubted attractions of IEZ as a way of dealing efficiently with the need to reconcile the demand for a good quality living environment with our present inability to prevent industry from creating nuisance beyond its physical confines.

Environmental zoning is not an entirely satisfactory solution, as it sanctions, at least temporarily, the continued existence of areas with substandard environmental quality, and places restrictions on the development of what may otherwise be prime land. Modern industrial societies are becoming decreasingly tolerant of industrial pollution, and more reluctant to accept the degradation of environmental quality caused by industry. In the longer term it seems likely that emissions that cause more than a negligible degradation of environmental quality outside the factory fence will not be allowed. This attitude will spawn new methods and technologies so that the effects of pollution and hazard will be eliminated. When and if this situation is achieved, there will be no further need for environmental zoning. In this view, the main role for environmental zoning thus is in the short and medium term, and new technology and attitudes will in the long term allow us to dispense with such an instrument.

Approaches for dealing with local environmental quality around industrial sites are still evolving. Many countries are beginning to develop a more integral approach to dealing with environmental problems, and this trend favours the emergence of IEZ-like schemes. It remains to be seen whether other countries will opt for such schemes, or will instead employ more pragmatic 'generic' buffering schemes as already observed in some countries.

Note

1   Jonathan Pearce is Freelance Advisor and Researcher for The Dutch Ministry of
         Housing, Spatial Planning and Environment.

References

Lloyd's Register (1993) *A Study on Environmental Zoning Systems in Twelve Industrialised
         Countries*, Report to Netherlands Ministry of Housing, Planning and Environment,
         January 1993.
Roo, G. de (1993) Environmental Zoning, The Dutch struggle towards integration, *European
         Planning Studies*, 1993-3.
VROM [Dutch Ministry of Housing, Spatial Planning and Environment] (1989) *Meerjaren Sa-
         neringsprogramma Geluidhinderbestrijding*, VROM, Leidschendam, The Netherlands.

Part D
Positive Environmental Spillovers in the Urban Areas:
A Neglected Field of Interest

Chapter 18

Seattle's Environmentally Critical Areas Policies

C. Marks[1]

18.1 Introduction

Seattle's Environmentally Critical Areas Policies and Regulations were adopted by the City Council in July 1992. They include geologic hazard areas (landslide-prone areas and liquefaction-prone areas), flood-prone areas, riparian corridors, wetlands, steep slopes, and fish and wildlife habitat areas. The goal of the policies is to regulate development within critical areas so that it occurs in a manner that is compatible with and appropriate to such areas, and protects the public health, safety, and welfare.

This chapter presents the background and history of this planning project, noting the reasons the project was undertaken and how the project changed after it began. The City's final policies were developed within the legal framework of Washington State constitutional law which places limits on the regulation of land use to avoid the 'taking' of private property. The policies were also developed within a public process involving competing and opposing interests. These policies were developed for a highly urbanised area, and the concept of externalities is central to the development of the policies. Finally, each type of critical area is summarised, as are the objectives and methods of regulating development in each of these areas. The need for effective implementation and enforcement of the regulations is noted.

18.2 Background and History

State Environmental Policy Act

The designation of environmentally sensitive areas in Seattle grew directly out of the State Environmental Policy Act (SEPA). SEPA was passed by the State legislature in 1971, and the City officially adopted its first SEPA regulations in 1978. SEPA establishes procedures whereby the City reviews the environmental impact of development projects.

SEPA review procedures allow certain exemptions from the environmental review process. These 'categorical exemptions' are for smaller projects, such as one or two dwelling units or very small commercial development, that generally are unlikely to have significant environmental impact. However, SEPA regulations recognise that some of these small projects can have impacts if the area within which they are located is very sensitive to development. In Seattle, almost all industrial, commercial, and residential developments must go through the SEPA process if they are located in a designated environmentally sensitive area.

In 1978, Seattle adopted the following environmentally sensitive areas in its SEPA ordinance:

- areas of steep slopes and potential landslide areas;
- flood-prone areas;
- areas extending landward for one hundred feet in all directions from designated creeks and lakes;
- areas subject to instability due to peat deposits and/or landfills;
- anadromous fish streams.

Problems with SEPA Maps

There were several perceived problems with the original sensitive areas maps. It was believed that they were not sufficiently complete or accurate. This was the main reason that the City undertook the effort in 1989 to improve the way it dealt with development in environmentally sensitive areas. It was believed that failure to identify areas as environmentally sensitive could lead to the inadequate review of a development application with resulting losses of property or lives, or degradation of the natural environment. Conversely, the designation of improperly identified environmentally sensitive areas could subject property owners to unnecessary environmental review, thereby inflicting costly and time-consuming permit review requirements upon them.

New Environmental Concerns

In addition to the mapping problems there were other reasons for undertaking this effort. Since the City's environmentally sensitive area maps were first prepared, a significant number of new environmental concerns had come to light which were not mapped. These concerns included the following: wetlands, abandoned solid waste landfills, toxic waste disposal sites, significant wildlife habitat areas, and earthquake hazard areas.

Development Pressure in Sensitive Areas

In the mid to late 1980s, Seattle saw increasing numbers of development applications in environmentally sensitive areas as more easily developed lots became scarce and as property buyers sought the scenic views that frequently accompany these sites. In the three years from 1984 through 1986 the City received building permit applications for an average of 279 dwelling units per year within designated environmentally sensitive areas. From 1987 through 1989 the annual average was 566 units.

Original Objectives of City Effort

In 1989 the City began an effort to improve the management of its environmentally sensitive areas. The original objectives of this City's effort were to:

- develop criteria for designating environmentally sensitive areas;
- propose policies for regulating development in these areas;
- develop regulations to implement the policies;
- conduct additional mapping of sensitive areas to improve the scope and clarity of the existing map.

18.3   How the Scope of the Original Project was Expanded

After the City began its work on environmentally sensitive areas in 1989 the project direction changed significantly with the effort taking on a much broader scope than originally envisioned.

Design with Nature

During the nineties many jurisdictions have begun to take natural conditions into greater account in their land use planning processes. Many of these jurisdictions identified areas with special characteristics that should be taken into consideration when reviewing development proposals.

The Seattle Planning Department set out to expand the scope of its sensitive areas work. This meant taking a more comprehensive view of regulating development in light of natural characteristics of various sites, rather than merely 'fixing' the sensitive areas maps.

Over thirty years ago, landscape architect and urban planner Ian McHarg wrote about *Design with Nature* (1969). This book was a seminal in regional and environmental planning. Lewis Mumford in the Introduction states, 'Man's life is bound up with the forces of nature, and that nature, so far from being opposed and conquered, must rather be treated as an ally and friend, whose ways must be understood, and whose counsel must be respected.'

McHarg says, 'Let us accept the proposition that nature is a process, that it is interacting, that it responds to laws, representing values and opportunities for human use with certain limitations and even prohibitions to certain of these'. McHarg proposed starting with a presumption for nature – to first identify areas most important to natural processes. These should then strongly influence the pattern of development. Critical land areas can be defined by their geologic, vegetative, or hydrologic characteristics. The appropriateness of an area for development can be determined by the importance of its natural resources, or the processes it accommodates, or the hazards it poses to urban development.

The idea is that development should generally occur where it will do as little damage as possible to the natural environment. Although McHarg's approach may be most relevant at a regional scale where large urban patterns are concerned, the idea that we should work with natural processes, and not against these processes, in planning for urban development is also applicable for development at a smaller scale.

The fundamental concept is that natural processes should be taken into account and development should occur in concert with these natural processes. Development practices should be designed in accord with rather than to overcome natural features of the land. As the adopted Environmentally Critical Areas Policies for Seattle states: 'Critical areas often present constraints to development that, if not recognised and mitigated, could cause damage to buildings and other structures such as bridges and utilities, and affect human health and safety. For example, flooding is a natural process and we should avoid placing buildings where they are likely to be damaged or destroyed by floods. Landsliding is also largely caused by natural processes, and likewise we should be extremely careful when developing landslide-prone areas.

In these and other areas, special care must be taken to protect the public health, safety, and welfare from land-related hazards, and to avoid damage to the ecosystem. Streams that serve as spawning grounds for salmon need to be kept free of silt and pollution in order to fulfil this role; wetlands serve many safety and environmental functions such as flood control, acting as a natural cleanser of polluted waters, and providing a rich habitat for wildlife.'

The section of the Critical Areas Ordinance titled Overall Goal and Implementation Principles further states: 'The Critical Areas Ordinance should allow land to be developed in accordance with the constraints and opportunities provided by the land itself. All land is the not same. If a person purchases a parcel that is 80 per cent wetland, it is significantly different than other types of property. The same is true with areas subject to landslides or floods. The owner purchased property that contained a wetland, or a landslide area, or a flood plain. The Critical Areas Ordinance should recognise that the reasonable development potential of such properties is less than the reasonable potential of unconstrained sites. The ordinance should permit development that makes use of a site's natural opportunities and that recognises its natural constraints.'

Washington State Growth Management Act

Another factor in expanding the original scope of Seattle's effort to develop critical areas policies and regulations was the passage of the State Growth Management Act in 1990.[2] The Act requires local jurisdictions to adopt a critical areas ordinance to protect these areas. 'Critical Areas' include the following areas and ecosystems:

- wetlands;
- areas with a critical recharging effect on aquifers used for potable water;
- fish and wildlife habitat conservation areas;
- frequently flooded areas;
- geologically hazardous areas.

Criteria, not Maps

One change in direction away from the original focus on 'correcting' the maps was the realisation that some of the source mapping was not originally intended to be extremely accurate. For example, because of the scale of geologic mapping that was used as the basis for the landslide hazard designations, the actual geologic conditions that cause landslides may be off by several tens or even hundreds of feet. However, over the years these maps had come to be seen as the final word concerning whether or not a site is subject to landslides. This was a misunderstanding of the original purpose of the maps which was to identify areas for further study.

Seattle's adopted Environmentally Critical Areas Policies and Regulations define areas based on specific criteria (for example, areas over 40 per cent slope), with maps serving only as general guides based on existing information. When specific actions are proposed in or adjacent to mapped critical areas, more detailed review may be required. Projects located within a mapped area could be exempted from the regulations if the applicant can demonstrate that the site does not, in fact, meet the definition of the critical area. Conversely, developments outside of the mapped area could be covered by the regulations if it is shown that the site does, in fact, meet the definition. Therefore, any lot within the city could be defined as a 'critical area' if it is characterised by the factors included in the applicable definition.

Development Standards

It became apparent that a major shortcoming of relying on SEPA review is that this review process does not use development standards, such as prohibiting the filling of wetlands or requiring a specific building setback from creeks. Instead, SEPA entails case-by-case review, which can result in mitigation measures or even in modified projects if the impacts are severe and unable to be mitigated. However, SEPA lacks the ability to regulate each development in the same and consistent manner, and must rely on individual project review to ensure that environmentally sensitive areas are protected. It is better to rely on development standards that are consistently applied in sensitive or critical areas. This became the basic approach used in the City's final critical areas policies and regulations.

18.4 Legal Framework

Seattle's critical areas regulations limit, in many cases, what a person can do with their land based on natural considerations. This is justified by local government's role in regulating development to protect the public health, safety, and welfare. However, there are legal limitations to this role. The U.S. Constitution states that a person's property cannot be taken without fair compensation and due process. The 'taking issue' is a very difficult one. It relates to 'reasonable use'. A person is not guaranteed maximum profit from a site but they do have a right to a 'reasonable' economic return.

In 1988 the Washington Supreme Court declared portions of Seattle's

Greenbelt Ordinance unconstitutional (Allingham versus Seattle). The Court said that the regulations that required a 'greenbelt preserve', in which no development was permitted, constituted a 'taking' without compensation and, therefore, was unconstitutional. Although the Court later reversed some of its findings in the Greenbelt case it didn't reverse its opinion that aspects of the ordinance were unconstitutional.

In a recent Washington Supreme Court decision (Presbytery of Seattle versus King County) the Court stated that there are two alternative ways to invalidate a zoning ordinance: 1) by establishing a 'taking'; and 2) on due process grounds. In order to withstand a 'taking' challenge, a regulation must meet two tests. First, it must protect the public from a harm (it must safeguard the public interest in health, safety, the environment, or fiscal integrity of an area). The ordinance should not go further and actually create a public good (such as preserving open space to be used by the general public). Secondly, the ordinance must not infringe upon a fundamental attribute of ownership (the right to possess, to exclude others, and to dispose of property).

Even if the regulation is insulated from a 'takings' challenge, it still must withstand the due process test of reasonableness. Here, there is a three-pronged test: 1) whether it is aimed at achieving a legitimate public purpose – it must prevent a public problem or evil; 2) whether it uses means that are reasonable and necessary to achieve that purpose – the regulation must solve the problem; and 3) whether it is unduly oppressive on the land owner. The third inquiry will usually be the most difficult and determinative one; it also lodges wide discretion with the courts. There are several factors that are often considered: the nature of the harm to be avoided; the availability and effectiveness of less drastic measures; and the economic loss suffered by the owner.

The Environmentally Critical Areas Policies and Regulations, under the Seattle City Law Department's guidance, were developed to be able to withstand the tests outlined above. Basically, a person must be able to achieve a reasonable amount of development. The basic concept is that there is a difference between preservation and protection. Preservation can be achieved through the purchase of land, all development can be prevented and public access established, and a 'public good' can be achieved. On the other hand, protection can be achieved through regulation designed to prevent a 'public harm', all development cannot be prevented, and an economic use of the property must remain.

The Seattle Law Department's concern with these issues and the recent court cases led to certain minor changes in direction in the development of the Environmentally Critical Areas Policies and Regulations. First, there was an effort to differentiate treatment of critical areas from open space issues and it was emphasised this was not intended as a replacement of the Greenbelt Ordinance. The concept of *Design with Nature* was somewhat de-emphasised; the issue was not whether land could be developed but how. An emphasis was placed on public safety issues and on preventing harm. Lastly, the critical areas policies treated all land with the same characteristic equally, in contrast to the failed Greenbelt Ordinance which regulated some tree-covered areas while somewhat arbitrarily not covering other similar lands.

18.5 Public Process

The Environmentally Critical Areas Policies and Regulations were developed within the framework of an extensive public participation process. An advisory committee was formed to guide the preparation of the policies. This committee was composed of representatives of neighbourhood groups, environmental organisations, developers, architects, geologists, and geo-technical engineers. Although this committee was not able to reach consensus on many issues, and while the meetings were divisive at times, the fact that this committee was involved was an important factor in gaining final City Council approval.

There was also an extensive public outreach effort. A series of seven public workshops was held to promote public involvement. Public participation was sought concerning the types of areas that should be addressed, and how the City could balance protection of critical areas with development objectives for privately-owned land. Over one hundred people attended these workshops. Valuable insight on public concerns was achieved through these meetings, especially regarding safety issues in landslide-prone areas and hillsides. While unanimity can never be achieved, this effort did lead to widespread public support for the overall critical areas policies which also was helpful in obtaining City Council approval.

18.6 Critical Areas in an Urbanised Context

Most jurisdictions dealing with environmentally critical areas possess large areas of undeveloped land. Seattle's situation is different since we have very little undeveloped land left. Our policies, therefore, focus on in-fill development. The fact that Seattle did adopt policies that place a large emphasis on natural conditions indicates that land-based ecological concerns can also be addressed in an urbanised area.

Some of our development standards reflect Seattle's urbanised environment. For example, we require only a 50-foot buffer from streams. While it could be argued that this is not sufficient to adequately protect water bodies from erosion and pollutants, it was our conclusion that this was the best that could be achieved. Any requirement of larger setbacks would not be practical given Seattle's small lot sizes, and it was reasoned that any natural buffer area is better than no buffer at all.

Table 18.1   Summary of environmentally critical area policies

Type of Area	Specific objectives	Identification	Implementation
Landslide-prone areas	• Ensure safe development • Prevent damage to neighbours	• Known/historic landslide areas • 15% slope and specified geological conditions • Areas over 40% slope • Previously altered steep slopes	• Project designed for 100 year project life expectancy and built to survive 1-in 100 year seismic event • Stabilise disturbed portion of site • Staged review permit process • Third party geo-technical review for hazardous situations • Control surface water drainage
Liquefaction-prone Areas	• Ensure seismically safe development	• Areas identified by USGS as subject to liquefaction	• Require geo-technical study • Require appropriate structural solutions
Flood-prone Areas	• Ensure safe development • Minimise flooding downstream • Manage surface water drainage system • Maintain water quality	• Areas on FEMA flood insurance maps	• No development in 'floodway' • Development in 'floodplain' should not increase down-stream flow • Change habitable space requirement from 1 foot to 2 feet above 100-year flood level
Riparian	• Control	• Class A Riparian	• Erosion and Drainage

Corridors	flooding and prevent property damage • Manage surface water drainage • Prevent erosion • Maintain/-enhance water quality • Maintain/-enhance fisheries and wildlife habitat • Maintain/enhance natural/native vegetation • Preserve quality of life and educational opportunities	Corridors: year-round or salmonid water bodies; includes Longfellow, Thornton, Pipers, Venema, Mohlendorph, Fauntleroy, Ravenna, Mapes, Dead Horse/Mill, Maple Leaf, and little Brook Creeks; Bitter and Haller Lakes • Class B Riparian Corridors: intermittent streams without salmonids • Plus 100 feet from water body or 100 year floodplain, whichever is greater, (for both Class A and Class B Riparian Corridors)	control on disturbed areas • Staged vegetation removal and replacement • No development or disturbance to occur within 50 foot minimum buffer (or 25 foot for intermittent streams) next to water bodies • Permit yard reduction, or buffer reduction, under certain conditions • Subdivisions -no development credit for buffer; up to full credit through conditional use • Subdivision configured to protect/incorporate water body and riparian corridor buffer • Early consultation with Department of Fisheries and Hydraulic Project Approval
Wetlands	• Prevent flooding • Manage surface water drainage • Maintain/ enhance surface water quality • Preserve fish and wildlife habitat • Preserve quality of life and education opportunities	• Areas inundated or satured enough to support a prevalence of vegetation typically adapted for life in satured soil • U.S. Fish and Wildlife Service inventory	• Erosion and Drainage control on disturbed areas • Staged vegetation removal and replacement • Prevent development on portion of site that is a wetland • Establish 50 foot minimum buffer around wetland where no development or disturbance may occur • Permit yard reduction, or buffer reduction, under certain conditions • Subdivision – no develop-ment credit for wetland/-buffer; up to full credit through conditional use • Subdivisions configured to protect/incorporate wetland and buffer • Require mitigation plans

Type of Area	Specific objectives	Identification	Implementation
Steep Slopes	• Minimise runoff and erosion • Maintain/ enhance water quality • Preserve tree, natural vegetation, and wildlife habitat • Maintain neighbourhood character	• Area over 40% slope	• Limit development on 40% slopes. For existing lots, develop least sensitive portion of parcel. • Subdivision – no development credit for steep slope area; up to full credit and/or possible intrusion onto steep slope through conditional use • Site disturbance limited vegetation removal and replanting • Site design guidelines: locate development on least sensitive portion of site, scale/design compatible with neighbourhood, encourage clustering, minimise terracing, preserve natural vegetation
Fish and Wildlife Habitat Areas	• Protect fish and wildlife habitat especially for 'priority species'	• Areas identified by Department of Wildlife as priority species habitat areas • All bodies of water that provide migration corridors and habitat for fish, especially salmonids	• Regulate development when also located in other critical areas to minimalise intrusion into habitat area, and protect against severing habitat corridors • Provide information for SEPA and shoreline program review to minimise and mitigate negative impacts
Abandoned Solid Waste Landfills	• Ensure safe development • Prevent subsidence and exposure to methane	• Sites listed in the *Atlas of Abandoned Solid Waste Landfills and Toxic Sites* (Source: Health Department) • Areas within 1000 feet of methane producing landfills • Sites identified by public or historical research	• Amend building code to prevent subsidence and require methane blocking membrane • Adopt Health Department recommendations for excavation and development of abandoned and closed landfill

Toxic Disposal Sites	• Prevent health hazards • Prevent further spread of contamination • Achieve cleanup	• Sites listed in the *Atlas of Abandoned Solid Waste Landfills and Toxic Sites* (Source: Health Department) • Sites discovered by historical research, site sampling, or during project review	• Amend Grading and Drainage Ordinance and related Director's Rule to ensure cleanup • Work with Health Department on sampling and cleanup plans

18.7   The Role of Externalities

Seattle's Environmentally Critical Areas Policies provide a good example of attempting to deal with externalities or spillover effects. For example, the landslide-prone area regulations specifically stipulate that development shall be strictly regulated to protect the public health, safety, and welfare on both the development site and neighbouring properties. Given the legal framework of avoiding regulating land that creates a public 'good' noted earlier, the policies do not emphasise the more generalised benefits that they can provide. However the policies do, in fact, have more generalised positive effects such as providing wildlife habitat and even aesthetic benefits that go beyond the borders of the regulated property itself.

A key aspect of the concept of externalities is that the free market often does not reflect all the social, environmental, or other costs that are involved with property development. This is central to the critical areas policies. For example, a development on hillside property could solve the landslide problem by paving over or removing much of the steep slope area. This would solve the landslide problem, but there would be no hillside left, and this action would not take into consideration erosion during construction, or the other environmental or aesthetic costs that doing this would entail.

18.8   Policies for Specific Kinds of Critical Areas

The policies dealing with each specific kind of critical area contain three aspects: 1) a general policy statement concerning the objectives of regulating development in these areas, 2) criteria for defining these areas and mapping them, and 3) implementation guidelines which have been incorporated into the City's land development regulations. Table 18.1 summarises these policies.

In some kinds of critical areas the type or extent of development is not regulated; the objective is only to ensure that development is carried out safely. For example, in landslide-prone and liquefaction-prone areas development may proceed as long as engineering solutions ensure safe development. In certain other areas, such as wetlands and their buffers and riparian corridor buffers, development is not permitted in order to protect the wetland or creek. Impacts in these areas cannot be prevented through engineering solutions and development restrictions are, therefore, provided.

Specific Policy Example: Landslide-Prone Areas

Following is an example of the three aspects (policy statement, identification criteria, and implementation guidelines) for the policy dealing with one type of critical area – landslide-prone areas.

Policy

Development on areas subject to landslides shall be strictly regulated in order to protect the public health, safety, and welfare on both the development site and neighbouring properties. The City shall ensure that engineering solutions are adequate to prevent failure during high stress periods and improper maintenance. The City shall ensure that public expenditures are not required to repair damaged facilities and protect against future damage due to instability created or exacerbated by development. The identification of landslide-prone areas shall include geologic, hydrologic, and topographic factors.

Identification Criteria Using Natural Processes

- *Topographic conditions* The slope of a site influences the tendency of near-surface soil deposits to slide. The greater the slope of the surface of the land, the greater the tendency for landslides; however, not all steep slopes are landslide-prone.
- *Hydrographic conditions* The ground or surface water environment on a site can adversely affect the stability of a geologic formation and provides the lubrication to cause soil masses to move.
- *Soils/geologic conditions* Slope stability in Seattle is strongly influenced by the physical character of underlying glacial deposits – the layering or stratigraphy of the soil or rock underlying the site. Many of Seattle's landslides are closely associated with contact between the Esperance Sand and either Lawton Clay or pre-Lawton sediments. This is illustrated in Figure 18.1. Water can readily move down through the Esperance Sand until it reaches the top of the Lawton Clay. Then the water moves laterally until it intersects a hillside. At this location there is often much seepage which contributes to ground saturation and can lead to landslides.

Implementation Guidelines

The City has adopted requirements for soil stabilisation and a review process that includes requirements for geo-technical studies and engineering standards whereby more restrictive requirements are imposed on more hazardous sites.

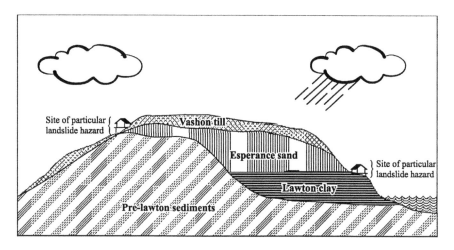

Figure 18.1 Idealised cross-section of a typical Seattle Hill
Source: Tubbs 1974

18.9   The Future: Need for Enforcement

The City recognises the need to translate general policies into regulations, to ensure that these regulations are correctly applied during project review, and that they are adequately enforced during actual construction. The City has hired a critical areas inspector to help ensure that these regulations are, in fact, applied correctly in the field. The City needs adequate follow-though from the plan review process, where conditions may be imposed on how development occurs, to the actual construction phase to ensure that these conditions are, in fact, carried out. For example, siltation fences may be required in steep slopes or near streams; it is important that these best management practices are instituted in the field.

Notes

1   Cliff Marks is working at the Seattle Planning Department.
2   With the passage of the Growth Management Act the City began referring to environmentally critical areas instead of environmentally sensitive areas.

References

McHarg, I. (1969) *Design with Nature*, Natural History Press, Doubleday and Company Inc., Garden City, NY.
Tubbs, D. (1974) Landslides in Seattle, Information Circular 52, Division of Geology and Earth Resources, Olympia, WA.

Chapter 19

A Wildlife Habitat Network for Community Planning using GIS Technology

K.J. Stenberg, K.O. Richter, D. McNamara and L. Vicknair[1]

19.1 Introduction

One of the greatest impacts of urbanisation on wildlife is habitat fragmentation. Parks programs and acquisition strategies can protect blocks of habitat. Ordinances that mandate the protection of streams and wetlands for public health and safety reasons also protect small bits of wildlife habitat. However, as these areas are surrounded by urbanisation and the landscape matrix shifts from natural vegetative cover types to urban land uses, they become remnant habitats that are disconnected and isolated from each other.

There has been much attention focused on the concept of corridors to connect habitat reserves. Generally this work has been done in non-urban areas (Noss 1987; Soule 1991; Noss and Harris 1986; Meffe and Carroll 1994) and is concerned with issues of which species are to benefit from the corridor and corridor width (Harrison 1992; Soule 1991). The predominant theories state that corridor widths should be at least as wide as the diameter of a hypothetically circular home range for the target species (Meffe and Carroll 1994). However, blocks of land that can accommodate the home range of an individual of a particular species begin to look a lot like small habitat islands rather than a corridor. What is needed is a way to connect those habitat islands and to provide a route through the urban landscape.

Urbanisation has occurred and will continue to occur in King County, Washington, therefore, we are attempting to create a system of nodes and linkages that will protect some habitat and wildlife values in the urbanising landscape. Through local regulations which protect streams and wetlands and open space acquisition programs, a tremendous amount of public money and effort is being expended to protect small to medium blocks of wildlife habitat. If we allow those areas to become isolated, we will lose many of the wildlife values that we are trying to preserve. Finding ways to connect those blocks of habitat could help to maintain their wildlife values.

In an environment of rapid urbanisation, development proposals are foreclosing options for wildlife protection almost daily. Unless linkages are identified prior to development, opportunities for wildlife habitat protection will be lost. This chapter will describe a method of identifying a habitat network in a rapidly urbanising planning area within King County and our success in implementing it through the planning processes.

19.2   Study Area

We focus on designating and assessing a wildlife habitat network for the East Sammamish Community Planning (ESCP) Area, one of thirteen sub-areas within King County. A comprehensive plan and area zoning for this sub-area was first adopted in 1982. The King County Council initiated an update in December 1989. Early in the process, loss of wildlife and wildlife habitats were identified as important issues by the citizens working on the plan update. Residents felt that the presence of wildlife and the natural vegetative cover contributed to the quality of life and 'sense-of-place' of the ESCP area.

The ESCP area is about 43 square miles in size and is about 15 miles east of the City of Seattle. The City of Redmond is located to the north and the City of Issaquah to the south of the planning area. The western edge of the planning area is defined by Lake Sammamish. The Sammamish Plateau, a glaciated drift plain that rises to about 500 feet elevation, makes up the majority of the planning area. The rolling topography of the plateau is dotted with lakes, marshes and bogs. The southern edge of the plateau is defined by Grand Ridge, a ridge of volcanic bedrock characterised by steep, rugged topography and five peaks, the highest rising to about 1400 feet elevation. The east and north edge of the planning area is defined by the Evans-Patterson Creek Valley. The southern boundary is located along Interstate 90 (I-90) and the Issaquah Creek East Fork Valley that separates Grand Ridge from Tiger Mountain to the south (Figure 19.1).

The ESCP area has been experiencing rapid growth. From 1980 to 1990 the area's census population has increased from 12,300 to 31,050. An analysis of the change in habitat values from 1981 to 1991 indicated that poor and moderate quality habitats increased 100 per cent. Excellent and good quality habitats that remain have become fragmented, isolated and much less available to wildlife. Average size of the higher quality habitats declined from about 51 acres to about 29 acres during that same time period.

19.3   Objectives

The objective of this study was to identify a network of wildlife habitats that would:

1   connect streams and wetlands;
2   form a continuous network which would provide some natural cover at build-out to allow animals to move across the landscape; and
3   connect the larger blocks of land in public ownership with other parts of the County.

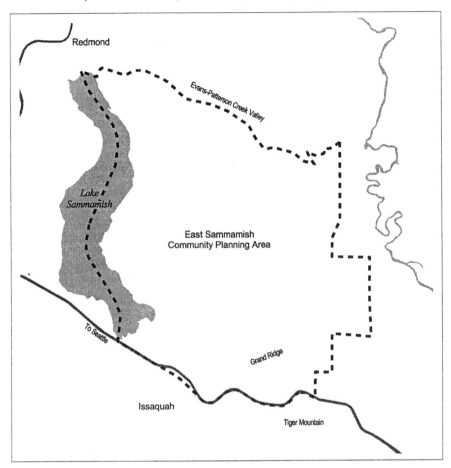

Figure 19.1   East Sammamish community planning area

These goals will help prevent genetic isolation and local extinctions of populations resident in habitat reserves.

While the analysis described in this chapter was done for a selected geographical area of the County, habitat reserves outside of the planning area were considered in the location of the network within the planning area. The intent was to expand the network Countrywide at an appropriate time. A countrywide habitat network has been identified and described in the 1994 King County Comprehensive Plan.

The habitat network will have a variety of incidental benefits to the human community. For example, the habitats in the network will contribute to groundwater infiltration, reduction of storm-water runoff, flooding and erosion, and visual relief from more urbanised landscapes. Through the community planning process, it was also determined that a portion of the network could serve as a community separator to delineate human communities that tend to lose their definition due to sprawl. The network may also provide residents with opportunities to view wildlife in their neighbourhoods.

19.4   Methods

In order to identify the best remaining locations for the wildlife network, high quality wildlife habitats were identified through analysis of natural resources and land use information in the County's geographic information system (GIS). The County uses a Hewlett Packard 720 workstation running Unix Version 7.2 on which ARC-Info 6.1.1 software is installed. Coverages used in the analysis included streams, wetlands, existing land uses, parcel size, and ownership size. We used a series of decision matrices to analyse the spatial information and rank habitat values. Areas that rank poorly on the final map would still support wildlife, however they are more likely to support a lower diversity of native species and a greater proportion of urban and exotic species. Areas that ranked highly in the final map were further investigated for inclusion in a habitat network. King County ranks wetlands and streams primarily on the basis of size, number of habitat types (complexity) and the presence of special species (Table 19.1). We used the County's stream and wetland rankings in the first part of the analysis. The regulatory buffer area for each stream and wetland class was included as stream or wetland area. Using the first decision matrix (Table 19.2), stream and wetland coverages were combined so that areas with high quality wetlands and streams received a high habitat value. Areas without streams or wetlands received a lower habitat value. The lowest habitat value is not equivalent to no habitat value for wildlife, it simply is lower on a relative scale when compared to the diversity of wildlife species that may be supported by wetlands and riparian corridors.

The land use map for the ESCP was generalised in terms of categories and was developed from aerial photo interpretation. The land use map was then digitised and the land use categories were collapsed into four classes that support differing levels of native wildlife species. The assumptions used to identify the four classes were based on research that relates wildlife diversity to housing density and type of open space (Stenberg 1988).

The second decision matrix (Table 19.3) builds on the results of the first. In the second step of the analysis we combined the streams/wetlands combination with the generalised land use map. The coverages were combined so that areas which were undeveloped and which had high quality streams and wetlands (which ranked high in the first step) were given a high habitat value in this step. Areas which were already developed, such as local shopping centres, and which did not have high quality streams and wetlands received a low habitat value.

Table 19.1   King County Wetland and stream classifications and regulatory buffers

Wetland Ratings	
Class 1:	Either greater than 10 acres, complex, or the presence of species of concern 100 feet buffer
Class 2:	Either between 1 to 10 acres, complex, forested, or presence of species of concern 50 foot buffer
Class 3:	Less than 1 acre and not complex 25 foot buffer
Stream Classifications	
Class 1:	Shorelines of the state; streams >= 20 cfs 100 foot buffer
Class 2:	Smaller than Class 1 Used by salmonids – 100 foot buffer Perennial stream, no salmonid use – 50 foot buffer
Class 3:	Intermitted or ephemeral; no salmonid use 25 foot buffer
Unclassified	Salmon use and/or perennial flow verified 25 foot buffer

Table 19.2   Decision matrix 1: wetland and stream combination

Wetland/Stream Combination				
Wetland Class → Stream Class ↓	Class 1	Class 2	Class 3	Class 4
Class 1	1	1	1	1
Class 2 with salmonids	1	1	1	2
Class 2 without salmonids	1	2	2	3
Class 3 and Unclassified	1	2	2	4
Not a Stream	1	2	3	5

For the third step we created two different maps from the parcel coverage. First, parcels were grouped into 5 size categories (Table 19.4). The size categories were determined partially by existing zones in the County's zoning code and partially by research which relates wildlife diversity to housing density (Stenberg 1988). This map was called the parcel size map. Second, the parcel coverage was 'dissolved' on ownership lines. The GIS allowed us to remove lot lines between adjacent parcels that are under one ownership. This 'ownership' map was then also grouped into the same 5 size categories as the parcel size map. This second map was called the ownership size map. The third matrix (Table 19.5) was then used to combine the parcel size map with the ownership size map. The coverages were combined such that areas with many small parcels that were already in separate ownerships received a low value.

Table 19.3   Decision matrix 2: land cover and wetland/stream combination

Wetland/Stream Value → Land Cover Value ↓	1	2	3	4	5
1. Natural Open Space, Undeveloped	1	1	1	1	2
2. Low Density Single Family	2	2	3	3	3
3. Modified Open Space	2	3	3	4	4
4. Medium to High Density, Multi-family, Commercial, Industrial, etc.	2	3	4	4	5

Resulting Values:

1   High quality streams and wetlands, high vegetative diversity, high quality habitats and natural cover.
2   Natural land cover, somewhat reduced diversity due to low impact residential, or absence of riparian systems.
3   Reduced vegetative diversity, moderate to low development density, or higher quality riparian systems found in higher density developments.
4   Low vegetative diversity, high level of development or poor quality riparian systems in moderate to low development densities.
5   Very low vegetative diversity, very high level of development.

Table 19.4   Parcel and ownership size classifications

Parcel Size Classification	
Acreage Range (acres)	Ownership size – Lot Size Value
0.00 – 1.00	7
1.01 – 2.50	6
2.51 – 5.00	5
5.01 – 10.00	4
10.01 – 50.00	3
50.01 – 100.00	2
Greater than 100.00	1

Very large parcels in one ownership received a high value. Areas still in one ownership but which were already subdivided to some degree received a medium value. This part of the analysis gave us an indication of the planning options that might be left in an area. If an area is already subdivided and in separate ownerships then planning tools such as clustering and setbacks are no longer options. In larger ownerships, creative site planning might still be applied to protect habitat connections.

The final decision matrix (Table 19.6) combined the parcel analysis with the results of the stream/wetland/land use analysis. This final combination resulted in the final habitat value map (Figure 19.2) which shows that there has been a lot of habitat loss on the Plateau. Approximately 50 per cent of the Plateau has experienced some level of development. High quality habitat areas have also become fragmented and isolated from one another and from the larger protected blocks of habitat that exist just outside of the planning area. However, the final habitat value map also showed us that there still was an opportunity to route a habitat network across the Plateau (Figure 19.2).

Table 19.5   Decision matrix 3: ownership and parcel size combination

Parcel Size Value → Ownership Size Value ↓	1	2	3	4	5	6	7
1	1	1	1	2	3	4	4
2	-	1	1	2	3	4	4
3	-	-	2	3	3	4	5
4	-	-	-	3	3	4	5
5	-	-	-	-	4	4	5
6	-	-	-	-	-	4	5
7	-	-	-	-	-	-	5

Resulting Values:
1 Most planning options left; good habitat value; low housing density.
2 Planning options slightly constrained; good habitat value.
3 Planning options moderately constrained; moderate habitat value; rural housing densities.
4 Smaller lots; few planning options; moderate habitat value.
5 Planning options extremely constrained; habitat values reduced; densities.

Table 19.6   Decision matrix 4: parcel size value and land cover/wetland/stream combination

Wetland/Stream Land Cover Value → Parcel Size Value ↓	1	2	3	4	5
1	1	1	2	2	4
2	1	1	2	3	4
3	2	2	3	3	5
4	3	3	4	4	5
5	3	4	4	5	5

Resulting Values:
1 *Best:* most diverse habitats; most planning options; largest lot sizes.
2 *Good:* most diverse to moderately diverse habitats; largest to moderate lot sizes – rural densities.
3 *Moderate:* most diverse habitats with smallest lots or moderately diverse habitats with moderately sized lots – rural densities.
4 *Fair:* good moderate habitat diversity with small lots op poor habitat diversity and largest lot sizes.
5 *Poor:* poor diversity on moderate to small lot sizes.
Categories 1, 2 and 3 were investigated further for potential network locations.

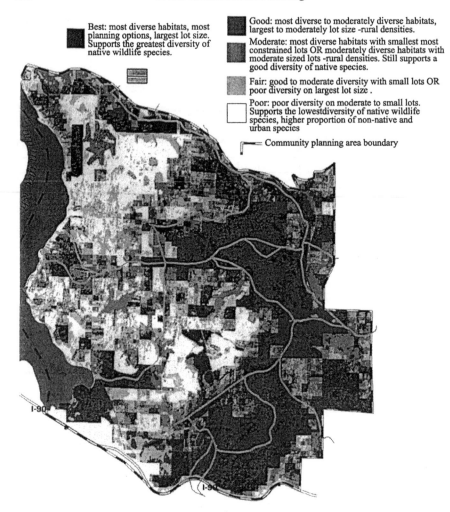

Best: most diverse habitats, most planning options, largest lot size. Supports the greatest diversity of native wildlife species.

Good: most diverse to moderately diverse habitats, largest to moderately lot size -rural densities.

Moderate: most diverse habitats with smallest most constrained lots OR moderately diverse habitats with moderate sized lots -rural densities. Still supports a good diversity of native species.

Fair: good to moderate diversity with small lots OR poor diversity on largest lot size .

Poor: poor diversity on moderate to small lots. Supports the lowestdiversity of native wildlife species, higher proportion of non-native and urban species

Community planning area boundary

Figure 19.2 Habitat value man and habitat network

To determine the actual network route, the stream and wetland coverages were overlaid on the final habitat value map. We returned to the stream and wetland coverages for network routing for both biological and political reasons. As animals move across the landscape, they will generally tend to follow riparian corridors. Riparian corridors frequently have a greater diversity of cover and the juxtaposition of water and upland attracts a wider diversity of wildlife species. In addition, current King County code requires buffers around streams and wetlands. By siting the wildlife habitat network in areas for which protection is already required, the network became more politically feasible.

Once a potential route was mapped out, it was field checked to ensure that natural cover was still present. A few areas were eliminated in this process. In an area experiencing rapid urbanisation, data only one or two years old can be inaccurate and ground truthing is necessary. The final route was entered into the GIS and display maps for public review were produced.

19.5 Results

The process identified a continuous network which connected the most important wetland systems and streams across the ESCP area. The network connects Lake Sammamish State Park, a large state park on the lakeshore, with other parts of the plateau. The final network crosses the plateau and extends from Lake Sammamish on the west to the major stream systems leading to the Snoqualmie River Valley on the east. It also connects Grand Ridge on the south to the Evans-Patterson Creek valley and stream systems on the north. The network should also help to maintain connections between natural systems on the plateau with the large blocks of land in public ownership to the south on Tiger Mountain.

There are two bridges on I-90 over the east fork of Issaquah Creek. Currently these are being used by wildlife as underpasses to cross from Tiger Mountain to Grand Ridge under the I-90 barrier. Trail plans propose using one crossing for human recreationists and leaving the other exclusively for wildlife. The network is designed to leave natural cover leading to and from these wildlife underpasses into the future.

The habitat network was adopted by the King County Council as part of the East Sammamish Community Plan in the spring of 1993. The area zoning for the community plan area was adopted at the same time (Figure 19.3). Lower than average housing densities are located along the habitat network where possible. An attempt was made to keep the average zoning at one unit per acre or less. Lower density zoning (1 unit per acre) through the centre of the plateau was also used to implement the urban separator concept.

Where the network is located along a stream or wetland, special zoning conditions were attached to the individual parcels which require additional buffer widths to accommodate wildlife. The standard buffers are not designed for wildlife protection and are too narrow for the purpose of a network. Where the network is not located in a stream or a wetland, such as in crossing drainage basin boundaries, the network width should be at least 300 feet wide.

Since much of the planning area is still undeveloped, it is too early to assess the success of the network in maintaining wildlife habitat values. As the area develops and the habitat network becomes the last remaining natural cover, this continuous habitat system will become more critical. It is also important to note that while the area may currently look undeveloped, many of the larger property owners have plans for residential developments. It may be possible to test the effectiveness of this network in the very near future. The only caveat is that the network must enter and exit the parcel where it is shown on the adopted network map. This will help ensure that a continuous network results as individual development proposals are prepared over time. We tried to give individual property owners as much flexibility in siting the network as possible. Often, on large parcels, many additional wetlands are discovered during the detailed site planning process. If a developer wishes to route the network in a different place within the parcel, in response to more detailed information discovered during site planning, that would be allowed.

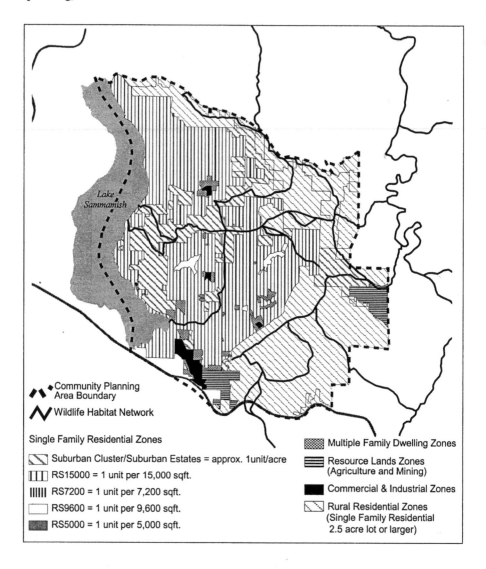

Figure 19.3   East Sammamish community plan area zoning habitat network

19.6 Conclusions

A wildlife habitat network was identified in the ESCP area based on a fairly simple model of habitat value. The model used information that was readily available in the County's GIS database. Since the type of information used is generally available to community planners, the method should be relatively easy to implement in other jurisdictions.

The network that was identified met the objectives of connecting valuable streams and wetlands, forming a continuous network across the planning area, and making connections to large blocks of public ownership outside of the planning area. Since much of the planning area is still undeveloped and covered with natural vegetation types, it is still too early to tell if the network will meet the additional objectives of preventing loss of habitat values in streams and wetlands and protecting species diversity in protected habitat reserves within the planning area.

As with all planning processes, the adoption of the network was a political process. Using the best information available we identified a network and then presented the proposed route to citizens, residents and elected officials for review and comment. The ESCP process involved over three years of public meetings and hearings. It is significant that the network concept received a tremendous amount of support from landowners, residents and elected officials. Towards the end of the process, there were two newspaper articles and two radio spots that focused solely on the network proposal. None of this media coverage generated any controversy. In addition, while the large developers in the planning area were not completely supportive of some aspects of the community plan, they were supportive of the habitat network. These large landowners even dubbed the network the 'emerald web'.

Some researchers have criticised the trend toward the use of wildlife networks or corridors (Simberloff and Cox 1987; Simberloff 1992). There is concern that corridors might actually enhance the spread of ecological disasters such as fires or of invasive, exotic species which out-compete native species. However, in an urban setting where the landscape matrix is already modified, connectivity between the remaining natural habitats declines.

Urban habitat reserves tend to be small, isolated and surrounded by a landscape matrix which supports invasive, exotic and urban species. A habitat network should consist of the original natural vegetative cover types and connect natural features and/or protected natural areas that were connected prior to urbanisation. Even if the habitat network does not achieve all of the objectives that we have set out for it, we agree with Noss (1987) that the original landscape was interconnected and to continue on a planning course that allows the landscape to become increasingly altered and disconnected without taking some action is imprudent.

Finally, it is important to note that a wildlife habitat network is only one piece of the wildlife protection puzzle. It will be necessary to use the full range of conservation tools, from acquisitions protecting habitat reserves to incentives and education which help people integrate wildlife into all of the places where we live, work and play.

Note

1   Dr. Kate J. Stenberg, Dr. Klaus O. Richter, Ms. Darcy McNamara, Ms. Lisa Vicknair
are working at the King County Environmental Division, Bellevue, Washington, USA.

References

Harrison, R.L. (1992) Toward a theory of inter-refuge corridor design, *Conservation Biology*,
Vol. 6, pp. 293-295.

Meffe, G.K., and C.R. Carroll (1994) *Principles of conservation biology*, Sinauer Associates
Inc., Sunderland, MA.

Noss, R.F. (1987) Corridors in real landscapes: A reply to Simberloff and Cox, *Conservation
Biology*, Vol. 1, pp. 159-164.

Noss, R.F., and L.D. Harris (1986) Nodes, networks, and MUMs: Preserving diversity at all
scales, *Environmental Management*, Vol. 10, pp. 299-309.

Simberloff, D., and J. Cox (1987) Consequences and costs of conservation corridors,
Conservation Biology, Vol. 1, pp. 63-71.

Simberloff, D., J.A. Farr, J. Cox, and D.W. Mehlman (1992) Movement corridors:
Conservation bargains or poor investments?, *Conservation Biology*, Vol. 6, pp. 493-
504.

Soule, M.E. (1991) Theory and strategy, in W.E. Hudson (ed.) *Landscape Linkages and
Biodiversity*, Island Press, Washington, DC.

Stenberg, K. (1988) *Urban macrostructure and wildlife distributions*, Ph.D. Dissertation,
University of Arizona, Tucson, AZ.

Acknowledgements

As with any local planning process, the East Sammamish Community Plan was the result of the
efforts of many people. The authors would like to acknowledge the contributions of the other
planning team members; in particular: Anne Knapp, Project Manager; Phil Dinsmore, GIS
Technician; Dennis Higgins, GIS Technician; and Kathy Creahan, Resource Planning GIS
Program Manager. We would also like to acknowledge and thank the residents of the East
Sammamish Plateau for not only demanding a new look at habitat protection in the Community
Plan but also for supporting the habitat network proposal through interminable public meetings
and hearings.

Chapter 20

Reducing Flood Damage Impacts of Urbanisation: Institutional Problems and Approaches

F. Westerlund[1]

20.1 Introduction

This chapter reports findings of a study completed at the University of Washington Department of Urban Design and Planning of needs for more effective flood hazard reduction planning in Washington State.[2] The main emphasis was on institutional issues; needs for better co-ordination of federal, state, and local efforts. However, there was a technical dimension as well, looking at new approaches to flood hazard reduction that are now being widely advocated, which may require new institutional mechanisms for their widespread adoption.

The relevance of this topic to the major theme of this book relates to the not new but still growing realisation that urban areas are not just the recipient of flood impacts; urbanisation can and does contribute to and exacerbate flood impacts on other urban and non-urban areas. These impacts are caused by many forms of physical development and modification of river channels, flood plains, and hydrologic basins, including building structures, fills, impervious surfaces, highways, bridges, and flood control structures themselves.

The vast Midwest floods of 1993 have focused attention on shortcomings of public policy related to flooding. In 1990, western Washington experienced unprecedented flooding in all the major flood plains and many tributary drainages and urbanising sub-basins. Many areas flooded that had never flooded before, including areas outside of mapped 100-year flood zones, even though peak flows in major streams were well below predicted 100-year event levels. The general conclusion was that urbanisation had a lot to do with this, particularly in causing localised flooding.

Some of the worst situations occurred in the Skagit River Delta north of the Seattle area, which is appropriate to mention here because it is the area of Washington often compared to The Netherlands, both because of its tulips and the fact that it is a diked flood plain virtually at sea level. Fir Island, a 70 square kilometer area, was under two to three meters of water for several weeks, resulting first from the failure of dikes and then by the retention of floodwaters behind those dikes. Cities on the Skagit River were protected by their levees, but flooded residents in unprotected communities on the other side of the river blamed their misfortune on these levees. Fills that have occurred in many places throughout the Skagit flood plain are regarded by neighbours as environmentally intrusive for their potential to divert and raise flood waters, yet filling in the flood plain continues. As a result of the 1990 floods, the Washington Legislature established a Joint Select Committee on Flood Damage Reduction to study the problem and recommend legislative action, co-chaired by a legislator whose family farm on Fir Island had been devastated, 'a lifelong believer in dikes and levees, until 1990'. Resulting proposed legislation, HB 1441 (Washington House of Representatives 1993), mandating county flood hazard reduction plans and a 'zero rise' state policy for flood plain management was unsuccessful in the 1993 legislative session, in part due to strong opposition from a few communities opposed to restrictions on flood plain development. As a parallel effort, the U.S. Federal Emergency Management Agency (FEMA) asked us to examine the institutional context for achieving flood hazard reduction, through better co-ordination of existing programs and new mechanisms if required.

Planning for flood hazard in the United States, like many other aspects of planning, is a fragmented patchwork of federal, state, and local authority. At the federal level, the Army Corps of Engineers has historically had a direct role in flood control on major rivers, through construction and repair of dams, levees, and dikes, or through programs to fund their repair after flood damage. The Corps and also state agencies regulate activities within or bordering waterways, such as dredging, filling, and levee construction. The state of Washington mandates local government land use planning and regulation within 200-foot wide shoreline zones under the Shoreline Management Act (SMA). The state Growth Management Act (GMA) requires identification of flood prone areas as a component of local land use planning. Both federal and state governments offer incentives such as low-cost flood insurance and funding, usually matching funding to local governments for planning and capital improvements related to flood hazard reduction.

Much of the responsibility for doing anything, either with new structural development such as levees or with non-structural approaches such as land use controls, now falls on local government; counties, cities, and also special taxing districts, of which diking districts are a common example in many areas. These local entities are confronted on the one hand by divided constituents, all in favour of flood hazard reduction but some opposed to land use restrictions of any kind, and on the other hand by a plethora of overlapping federal and state programs that could assist flood hazard reduction. Most of these programs are voluntary. A few such as SMA and GMA impose loose requirements for flood-related land use planning, but with few specific guidelines. All of these programs are complex, time-consuming, and fraught with paperwork.

20.2   Study Approach and Findings

We reviewed documentation for eight federal and state programs concerned with flood plain management (which is the traditional term for flood hazard reduction planning), including FEMA's National Flood Insurance Program, Army Corps of Engineers construction and maintenance programs, and federal and state programs which fund local governments to do flood hazard mitigation. About 35 people were interviewed, including managers of these programs, local government officials, planners and their consultants, state legislators, and citizen activists.[3] The majority of those interviewed, including program managers who one might expect to be defenders of the status quo, were proponents of a major paradigm shift from structural to non-structural approaches. Virtually everyone acknowledged that flood control based on levee and dike systems and channel improvements has reached its limits, that little expansion of these systems is likely or could be justified as cost-effective, and that these systems have not reduced long-term risk for many of the areas they were intended to protect and have been detrimental to other riverain resources such as fish habitat.

The Corps of Engineers has proposed a plan to replace existing levees in the Skagit Delta with levees designed to overflow or overtop at flood tolerant locations, to reduce peak flows in the Skagit River. Others interviewed spoke of lowering, removal, or setback of levees to increase storage and conveyance of flood-waters. King County, through its Surface Water Management Division, has an active flood hazard planning program that includes these elements (King County Surface Water Management Division 1993). Snohomish County has been working with its diking districts to get them to co-ordinate their dike systems, lowering many of them to a common level that equitably distributes flood impacts.

The need to restrict development in flood plains is not a new idea. It was acknowledged by everyone except for a few local officials in small communities located almost entirely within flood plains. Here the argument was for flexibility that recognises the particular situation of a community in relation to its geography. They were saying, in effect, give us some options and do not hold us to an absolute, rigid standard.

The National Flood Insurance Program (NFIP), administered by FEMA, has since 1968 been the major institutional response at the federal level to the problem of flood plain development. NFIP is a voluntary program allowing property owners in participating communities to obtain affordable, federally-underwritten flood insurance, if the community adopts a flood plain ordinance requiring new residential development to be elevated above the 100-year flood level, and other development to be either raised or flood-proofed. Development is also prohibited in floodways, unless built in the downstream shadow of other development. The floodway, the area to be kept free of obstruction, is defined by a one-foot rise model (Figure 20.1), that assumes obstruction of floodwaters by either levees, filling, or other development in the flood plain to the extent that produces a one-foot rise of the water level in the stream channel above the base flood elevation.

In its adopted local ordinance the participating community is not supposed to permit any development that would cause more than a one-foot rise when the cumulative impact of that development along with existing and expected future development is estimated. However that is a local government determination which is rarely made because cumulative impact analysis is a difficult. In reality, according to most of those interviewed, each development is considered separately, and since no one development can cause a significant rise throughout the flood plain, nothing gets denied as a result of this provision. Over time, entire flood plains can be filled and developed.

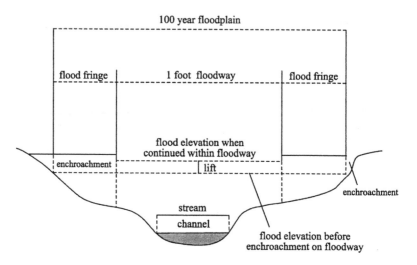

Figure 20.1   FEMA one-foot rise floodway model

Evaluation of the NFIP was not our central purpose. The program has its critics; the reader is referred to a paper by Bollens, Kaiser and Burby (1988) of the University of North Carolina Planning Department who argue that the program has actually encouraged flood plain development. A zero-rise standard (Figure 20.2) would be much more restrictive of flood plain development. This is often misinterpreted as a 'zero-development' policy, though in fact, significant development could occur in flood plains if its impacts were mitigated, by providing compensatory storage elsewhere in the floodplain, building on existing fills, or setback of existing levees to increase conveyance area. New levees could even be constructed with a zero-rise floodway by providing compensatory conveyance (Figure 20.3).

Another concept is called a density flood plain. It would be based on some maximum allowable increase in water surface elevation, perhaps less than one foot, but this would be calculated based on a maximum percentage of the flood plain's area that could be filled. That percentage could be applied to each property; it might be 10 per cent of the area of each property that could be filled for development, or a sliding scale of buildable percentage based on lot or parcel size, with a larger percentage for small lots. Also, any open space set-asides would figure into the equation, allowing a larger buildable percentage for the remaining area.

Since a density flood plain concept would be implemented through zoning that is also directed at other planning objectives, it offers an example of an 'integrated environmental zoning' approach, and one that is based on impact modelling.

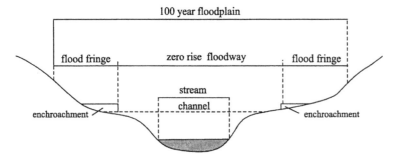

Figure 20.2 Zero-rise floodway model

The FEMA 100-year flood plain, as a single designation for flood prone area, was also seen as limiting. It was suggested that different levels of flood hazard exist, which could be defined say, by 10, 25, 50 and 100-year flood levels, with different degrees of regulation or mitigation for each zone. Examples of mitigation include government buyout and removal of structures in the 1-10 year zone, flood proofing of existing development and prohibition of new development in the 10-25 year zone, and lesser restrictions or regulation based on density flood plain concepts in the 25-50 and 50-100 year zones. Little sense was seen in requiring the same restrictions in areas of 100-year frequency that apply to areas of 10-year frequency.

Because most flood prone communities participate in NFIP, it has been accepted as a minimum standard for flood hazard planning almost everywhere, but the minimum standard tends to become the only standard. Only eight states and scattered local governments have adopted more stringent standards, and often these relate to minor particulars. For example, the Washington Department of Ecology, which largely administers the NFIP within the state, does not permit downstream shadow development in floodways, but only in this respect are Washington's standards more stringent.

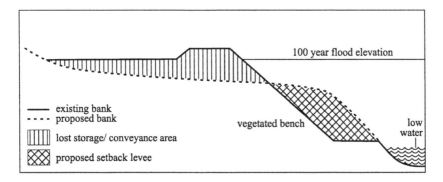

Figure 20.3 Levee construction consistent with a zero-rise model
Source: King County Surface Water Management Division 1993

FEMA officials acknowledge that NFIP by itself is not a planning program, has limited objectives, and needs to be combined with other efforts and integrated into local planning. A recent add-on to it is the Community Rating System, which allows a community to qualify for reduced insurance rates if it adopts certain other measures, such as a flood hazard element in its comprehensive plan, a public education program, or a flood warning system. However, to qualify, communities have to apply separately for this program and deal with another set of program requirements; few have so far.

Another FEMA program is the Hazard Mitigation Grant Program which provides matching funds for mitigation efforts following a flood. This has mainly been applied to rebuilding flood control structures that have failed or suffered damage, and in the absence of other local policy direction is limited to this kind of assistance. But FEMA officials emphasised that it can be applied to new forms of mitigation, including non-structural approaches, if such are clearly called for in an existing local plan.

At the state level, the Department of Ecology also provides matching funds to localities for flood hazard mitigation that usually involves but is not limited to structural approaches. In this case local governments must have a plan, and currently this must be a plan prepared specifically for that program; local governments cannot simply point to a flood element in their comprehensive plan. Such separate requirements may actually discourage the development of flood hazard reduction elements in comprehensive plans, where there can be effective integration with general land use planning.

A number of other issues surfaced in the interviews. A major one relates to the relationship between flood plain management and watershed management. These tend to be separate worlds in most cases. Flood plain managers observe that land use practices upstream, such as forestry practices in upper basins and urbanisation in smaller sub-basins tributary to flood plains, have increasing impact on floods, but they lack tools in their programs to deal with these matters. It is largely a set of jurisdictional problems. Basins extend across county lines; land in upper basins may be managed predominantly by federal or state agencies, or timber companies under state or federal regulation. Urbanisation may be occurring in incorporated cities while major flood plains downstream are in the unincorporated county.

Even within one government and one agency there may be separate management regimes. King County's Surface Water Management Division (SWM) does both flood hazard reduction planning for the major rivers and flood plains and urban runoff analysis for smaller basins and sub-basins, the latter for purposes of stream protection, review of drainage plans and design of detention systems for proposed development. Both activities involve use of hydrologic models; runoff models that predict flows (flood hydrographs) during a storm event, and flood plain models that generate cross sectional profiles, predict water surface elevations, and areas of inundation. These models can be linked to do flood analysis on a basin-wide or watershed level, and SWM is moving in this direction, but until now these have been organised as separate activities. Most interviewees felt that a basin-wide approach will ultimately need to be taken for every major river basin that has flood hazard, employing basin-wide modelling. Who will do this modelling and how it will be paid for are unanswered questions.

Another area touched on is that of public and private liability for flood impacts. Structural solutions prevent flood damage in certain areas and may cause it in others. Public agencies must accept responsibility and liability to equitably protect lives and property in flood prone areas, or they may be forced to in the courts. Such court cases have begun to appear. New legal mechanisms may be needed to solve some of the equity and property rights issues involved in flood damage strategies, including impacts of permitted private development actions on other property owners.

The state of Wisconsin offers an interesting example. Flood plain development is permitted, but anyone who proposes a development that would cause a 0.1-foot rise in floodwaters anywhere in that reach of the river, upstream or downstream of the proposed development, must purchase flood easements from affected property owners (WDNR 1992).[4]

The interviews served the purpose of identifying this broad set of problems and issues related to flood hazard reduction, and also indicated that wide areas of agreement apparently existed among key members of the flood management and planning community in Washington. We therefore convened two full-day roundtable meetings of the group we had interviewed in an effort to generate ideas and encourage agreement on recommendations for improved program co-ordination that would promote some of these new approaches. Attendees included two state legislators (including the sponsor of HB 1441, who planned to introduce revised legislation) and representatives of the state Department of Community Development, which oversees the Growth Management Act, the legislation that requires most localities to do comprehensive planning.

The meetings were quite productive. The people involved in the various flood plain management programs had all dealt with each other and seemed to form a tight-knit group, yet had never all met together in one place. Most had very little knowledge of the Growth Management Act and its objectives for co-ordinated land use planning. Out of these meetings came a strategy for integrating all of the existing programs into GMA comprehensive planning at the county level in a way that would satisfy the requirements of each existing program and at the same time support and encourage some of the creative new approaches that had been identified. This strategy is presented in the recommendations which follow.

20.3   Study Recommendations

The State of Washington should recognise that floods are natural events that cannot be controlled without altering natural systems of flood plains and watersheds in ways that are detrimental to our riverain resources, and which obligate future generations to pay expensive maintenance costs for systems that historically have not reduced long-term risk.

Therefore, it should be fundamental policy to reduce risk and potential flood damage by avoiding significant alteration of the hydrologic processes of flood plains and watersheds. Plans and actions regarding land use and development must respect flooding as a natural process and should be designed to avoid or mitigate impact on flooding.

Counties are the appropriate unit of government for the development and implementation of flood hazard management plans. However, not all counties in Washington have significant flood hazard. A method should be developed for designating target counties. One concept would be to designate counties that have had three federal declarations for flood hazard. Designated counties should be required to prepare a flood hazard management plan that is either county-wide, or basin-wide for each major watershed that has identified flood hazard. Such plans should address entire watersheds as well as entire flood plains within them, and should give attention to forest practices in upper watersheds and development practices in urbanising sub-basins, as well as land use and flood hazard reduction in downstream flood plains.

Flood hazard management plans should become an element of comprehensive plans developed under the Washington Growth Management Act (GMA). Planning for flood-prone lands is already a GMA requirement. If requirements for flood hazard planning from other state and federal programs are incorporated in flood hazard plans prepared as a part of GMA comprehensive planning, this will further encourage localities to deal in a critical way with the land use/flood hazard relationship. GMA would thereby provide a framework for co-ordinating flood hazard planning that is now being conducted under several separate state and federal programs, including FEMA's National Flood Insurance Program (NFIP), Community Rating System (CRS), and Section 409 Hazard Mitigation Grant Program (HMGP), and the Department of Ecology's Flood Control Assistance Account Program (FCAAP).

GMA consistency requirements are of key importance in establishing co-ordination. City plans and actions must be consistent with the county plan, and state agency actions must respect these plans. All capital programs must be consistent with comprehensive plans. The relationship between GMA and the State Environmental Policy Act (SEPA) is significant. Programmatic environmental impact statements under SEPA for GMA comprehensive plans that incorporate flood hazard planning would need to consider hydrologic effects at full build-out, requiring assessment of cumulative, watershed-scale impacts.

County flood hazard management plans prepared under GMA should follow current or expanded guidelines developed by the Department of Ecology for Comprehensive Flood Control Management Plans (CFCMP) required under FCAAP. County flood plans should become an annex to the Washington State Flood Damage Reduction Plan (prepared under Section 409, Stafford Act). Cities within designated counties should also develop hazard management plans or be appropriately included within the county plan.

If the above conditions are met, and a community is participating in the FEMA National Flood Insurance Program, then the following conditions should be accepted: (1) FEMA acknowledges the community's flood plan as prepared under GMA as meeting requirements of the NFIP and CRS programs and, should the community elect to participate in CRS, it should qualify for a reduction in insurance costs; and (2) FEMA Section 404 and DOE FCAAP funds could be used in co-ordination to implement the community's flood management plan.

The State's match for permanent work under the Stafford Act should be restricted or denied for non-complying communities; those which do not prepare plans, or whose plans are deficient in meeting state standards, or which do not meet an established timetable. A grading or rating system on a sliding scale as a basis for rewarding funds might be instituted as an incentive mechanism.

The existence of a local plan is particularly significant for FEMA funding following a flood. Under Executive Order 1198, FEMA can rebuild or fund 90 per cent of loss value to implement other mitigation if it is contained in a local plan. This includes adhering to development regulations and building codes contained in or enacted to implement a local plan.

Because of the important relationships between GMA, DOE-FCAAP, the State Flood Damage Reduction Plan, and the disaster response role of the DCD Division of Emergency Services, a co-ordinating mechanism at the state level is needed. It is recommended that a permanent Flood Damage Reduction Board be established as envisioned in the State Flood Hazard Mitigation Plan, with responsibility for co-ordinating the involvement of these state programs in flood hazard management planning.

Counties and cities should have the flexibility to devise local solutions, not constrained by highly specific requirements imposed from above. This is consistent with the 'bottom-up' architecture of the Growth Management Act, and the fact that land use policies are a local prerogative. Standards such as 'zero-rise' or 'no net fill' should be seen as tools available to communities that wish to use them, but they do not fit all situations and should not be imposed as requirements. However, communities must recognise the state and national interest in flood hazard reduction. Their interests are expressed in the identification of target counties and watersheds, the requirement that watershed-level flood hazard management planning be conducted and implemented through land use regulations, mitigation, and capital programs, and the principle that flood impacts cannot be imposed on neighbouring jurisdictions. Minimum state guidelines should include the requirement that flood plans: (1) identify risks to lives and property, (2) identify potential damage and how to reduce it, (3) identify impacts on other natural resources, and (4) show how the plan if implemented will reduce long-term public costs and not cost more than potential flood losses. Promising approaches for flood hazard reduction include:

1   encouraging or requiring flood-tolerant land uses in flood plains that can accommodate periodic overflows;

2   establishing flood plain zones for flood frequencies less than 100 years (e.g. 10, 25 or 50 years), with different land use controls or mitigation requirements for each;

3   levees and dikes designed to overtop at certain flood-tolerant locations during peak flows of major floods. This can involve removing, lowering, or setback of existing levees, as well as selected new construction;

4   co-ordination among levee and dike systems throughout a river basin so that flood impacts are equitably distributed;

5   employing computer-based hydrologic models developed specifically for each

basin, such as the model the U.S. Army Corps of Engineers has proposed to develop for the Skagit River basin. Analysis using such models would make it possible for planning to be tailored to the physical characteristics and existing/proposed development conditions of each basin. The State can assist in such efforts by developing models, providing GIS capability, and assembling watershed characteristics information;

6   low-cost, cost-effective flood warning and public education programs, such as that developed in Lewis County, a locally implemented effort assisted by the Corps of Engineers.

The Washington Department of Transportation (WDOT) is a major player in development statewide and should be a participant in the formulation of state policies for flood hazard reduction. Bridges and highways are major factors in flood dynamics, and new construction as well as replacement of existing structures should give full consideration to flood-related effects.

Structural solutions such as levees prevent flood damage in certain areas and may cause it in others. Public agencies must accept responsibility and liability to protect property in flood prone areas. New legal mechanisms may be needed to resolve the equity and property rights issues involved in flood damage reduction strategies; such issues as how a property owner's rights are affected by regulation of development, and by permitted development actions of other property owners. Washington may be able to benefit from the experience of other states that have addressed this issue.

There is a need to deal compassionately and responsibly with worst-case situations, especially those that involve individuals who own property and reside in floodways or flood-prone areas of flood plains and whose damages exceed 50 per cent of value, thus preventing rebuilding or financial recovery under present rules. FEMA is seeking legislation that would amend the National Flood Insurance Program by providing a relocation option in such circumstances. The development community should be receiving greater assistance and guidance in building projects which are sensitive to flood hazard avoidance. Education and professional assistance needs to be designed and directed to developers, as well as local government regulators, to transfer knowledge about flood hazard reduction methods and their costs.

This summarises the major findings and recommendations of the study. Some more specific recommendations followed regarding actions that could be implemented by agencies without passage of new state legislation, and actions for which new legislation would be needed. The former included the development, jointly by the state Department of Ecology and Department of Community Development, of a guidebook for the preparation of county flood hazard reduction plan elements, based on adaptation of DOE's existing guidebook for FCAAP Flood Control Management Plans and GMA critical areas planning guidelines. New legislation would be required to officially identify flood prone communities, establish state goals, objectives, and standards for flood-planning, and institute restrictions on state funding for non-complying communities.

To insure basinwide natural resource planning and integration of flood impact assessment within GMA comprehensive planning, the need for possible amendments to the Growth Management Act and State Environmental Policy Act was also envisioned.

Although much of this final work was specific to the situation in Washington State, a similar framework for flood hazard reduction planning and a similar set of issues would be found in most regions of the U.S., in large part due to the common federal role and its history. States have gone in quite different directions, however, with respect to the adoption or non-adoption of state-mandated land use planning and growth management, so the lessons of this study may have useful application in some places and not others.

Notes

1   Frank Westerlund, Ph.D. is Associate Professor at the Department of Urban Design and Planning, University of Washington, Seattle.

2   The study was supported by a grant from the Federal Emergency Management Agency (FEMA) through the Growth Management Planning and Research Clearinghouse, Department of Urban Design and Planning, University of Washington. In addition to the author, participants included Dan Carlson, Associate Director of the Clearinghouse, and graduate students Jeff Henderson and Liza Joffrion.

3   For the rather extensive list of agencies and individuals interviewed and later participating in the roundtable meetings, the reader is referred to the study's final report (Carlson and Westerlund 1993).

4   Information about the Wisconsin program was provided by Mr. David K. Carlton of KCM Inc., Seattle, who formerly worked for the Wisconsin Department of Natural Resources in developing that state's program.

References

Bollens, S.A., E.J. Kaiser and R.J. Burby (1988) Evaluating the Effects of Local Floodplain Management Policies on Property Owner Behavior, *Environmental Management*, Vol. 12, No. 3, March 1988, pp. 311-325.

Carlson, D., and F. Westerlund (1993) *Flood Damage Reduction in Washington*, Growth Management Planning and Research Clearinghouse, Department of Urban Design and Planning, University of Washington, Seattle, WA.

King County Surface Water Management Division (1993) *King County Surface Flood Hazard Reduction Plan*, King County Department of Public Works, Seattle, WA.

Washington House of Representatives (1993) Committee on Environmental Affairs, Engrossed Substitute House Bill 1441, *An Act Relating to Flood Damage Reduction*, Washington House of Representatives, Olympia, WA.

Wisconsin Department of Natural Resources (1992) *Floodplain and Shoreline Management, A Guide for Local Zoning Officials*, Wisconsin DNR, Madison, WI.

Part E
The Spatial Plan as a Tool for Sustainable Solutions

Chapter 21

Environmental Appraisal of Development Plans

N. Border[1]

21.1 Introduction

The process of environmental appraisal discussed in this chapter was developed in the context of the British planning system and relates to current British legislation and guidance. However, the process is flexible and can be adapted to other planning systems, making allowances for different practices and regulations.

This chapter draws upon a research study which operated from December 1992 to June 1993. The Research was carried out by a joint team from the University of the West of England and Baker Associates, Planning Consultants. The core research team comprised John Baker, Hugh Barton, Neil Border, Sally Downs and Peter Fidler.

21.2 Pressures on Urban Planning

Urban planning faces a conflict between positive environmental policy and effective land use control. There is pressure to promote more compact developments, reducing capital energy and resource inputs, minimising travel distances and creating opportunities for the use of a wide range of modes of travel other than the car. Yet there is a need to encourage and plan for economic growth – possibly involving sources of disturbance or pollution, such as production centres or power stations. This growth must be accommodated in such a way as to allow the public to access facilities and employment opportunities; private industry to maximise production and profits; yet with a minimum of disturbance to the environment and established land uses, such as housing and schools.

This act of balancing economic growth, social issues and environmental protection is not new; all planning systems have been fighting this battle for a long time. Now, however, there is a new political and social emphasis on environmental protection. Economic growth and social issues are to be considered in light of their possible impacts on the environment, both locally and globally. The balancing act has started to become a conflict, with economic growth and social issues on one side and the environment on the other.

Since the Rio Earth Summit national governments have been stating their intentions to work towards the objectives of sustainable development and growth

(British Government 1994). In dealing with land and its use, planning and in particular the preparation of development plans, can contribute to these objectives. Embracing the concepts of sustainable development and growth will go some considerable way to resolving the conflicts now facing urban planning.

It is worth noting that through the integration of sustainable development, land use planning will not only move towards the resolution of the conflicts it now faces, but will also contribute to attaining the goal of sustainability (Breheny 1992). This research took sustainability as meaning 'that the environment should be protected in such a condition and to such a degree that environmental capacities (the ability of the environment to performs its various functions) are maintained over time: at least at levels sufficient to avoid future catastrophe, and at most at levels which give future generations the opportunity to enjoy an equal measure of environmental consumption' (Jacobs 1991).

The key definition of sustainable development used in the research was given by the Brundtland Report and states 'In essence, sustainable development is a process of change in which the exploitation of resources, the direction of investments, the orientation of technological development, and institutional change are all in harmony and embrace both current and future potential to meet human needs and aspirations' (United Nations Commission on Environment and Development 1987).

21.3   Conflict Proving Difficult to Resolve

It is clear why this conflict between economy, social issues and environment is proving difficult to resolve. Firstly, there is the problem of identifying and analysing the effects of development plans and policies. Secondly, it is not clear which issues should have priority and whether the planning system should be weighted in favour of one particular set of considerations.

To deal with the second problem, the nature of the planning system must be understood. It is, on the whole, a political activity where decision-making is undertaken by a committee of elected members, acting in the interest of the public. The weighing of issues is therefore to be determined as part of the political process – the final decisions on whether to favour society, the economy or the environment are in the hands of the elected members.

The former problem – identifying and analysing the effects of development plans and policies – is closely related to the second in that it is this identification and analysis which contributes to the decision-making process. Information and advice gathered during this analysis can be used by the decision-makers when considering the proposals and policies of a plan and deciding upon the weighing to be given to each issue.

The importance of analysing the potential and actual impacts of policies and proposals, and providing the most accurate information to the decision-makers, demands a methical system that can be integrated into the established planning process. In this way, the identification and analysis of environmental impacts can be undertaken as an element of preparing development plans (see Department of the Environment 1992).

21.4 Environmental Appraisal of Development Plans

To appraise the potential and actual impacts on the environment of a development plan means to add a new dimension to the preparation of a development plan. Potential impacts are relevant for new policies or proposals where the impacts are only in the forecast stage, whereas actual impacts are relevant where the policy is current and the effects can be identified and monitored. Each policy and proposal contained within a development plan must be considered not only in terms of its influence on economic conditions and the interactions of society but also in terms of its environmental impact. This extends beyond what might be considered to be 'environmental' policies, to include policies on housing, recreation, transport, employment and so on.

The term environmental appraisal means an explicit, systematic and iterative review of development plan policies and proposals to evaluate their impact on the environment (see Commission of the European Communities 1991). As such it is an integral part of the plan making and review process.

In this wider consideration, the environment to be addressed is not merely that found within the local authority boundary, but also the national and international environment. Policies and proposals need to be seen in light of their effect on the environment as a whole, from the local to the global, including such issues as carbon dioxide emissions, tree cover and energy use.

21.5 Methodology

In preparing the guidelines, the study concentrated on practical, achievable and cost-effective approaches and practices which could be taken forward, in the context of available resources and knowledge, and developed for the environmental appraisal of development plans.

The study was informed by a wide-ranging review of relevant literature in the fast-growing field of environment and sustainability. There was an in-depth examination of literature and other source material exploring methods and techniques of appraisal, assessment and audit in planning and related practice. The primary empirical work for the study involved a number of strands. These were:

- a comprehensive survey of all local planning authorities in England and Wales;
- consultations with a wide range of other key bodies and interests;
- detailed case studies of ten selected local planning authorities undertaking environmental appraisal in a range of development plan settings.

As the study progressed, the ten case studies were enhanced by follow-up work on two supplementary local authority reviews where further good practice was identified.

21.6 Environmental Appraisal Guidelines

The research resulted in a set of guidelines (Department of the Environment 1993) aimed at enabling every plan-making authority to undertake environmental appraisal, in a way which contributes to the preparation of good development plans. The appraisal process proposed is intended to be adaptable to every level of plan and government, and to the skills an authority either has within its plan-making team or can realistically call upon.

Key Points

The guidelines identify a number of issues which must be embraced by environmental appraisal if it is to be effective and efficient. These include points pertinent to conducting an appraisal as well as more general concepts for contemplation prior to appraisal. The study identified the following key points:

- environmental appraisal applies to those authorities setting out on the process of plan preparation or review, and for all authorities in subsequent rounds of plan-making. It is possible to conduct environmental appraisal at any stage in plan-making – even a brief appraisal which may seem simple in scope will still afford benefits in terms of improved plan quality;
- appraisal must ensure that the scope of the plan covers the appropriate range of environmental concerns in order to guarantee consideration of appropriate options and to prevent omissions;
- appraisal must assess policies and proposals to establish their environmental effects. This is an iterative task which involves refinement, improvement and, if appropriate, development of ancillary policies or proposals aimed at mitigation; it will take place at several stages of plan-making;
- as a process environmental appraisal requires that the current baseline be established – the state of the authority's environment at that time (see Beanlands 1990; Clark and Herrington 1988);
- the appraisal itself should be well structured and systematic, yet avoid unnecessary complexity. Structured to secure a transparency of the appraisal, making it easy to follow and understand. Systematic to ensure that the impact of each policy or proposal is considered in terms of the current baseline;
- the appraisal should be conducted in house whilst the development plan is in preparation, as an integral part of that preparation process, not a retrospective supplement;
- the appraisal process must be flexible to enable its use at all levels of government, reflecting different degrees of resource availability, thereby providing comparability and consistency between planning authorities, both at the same planning level and at different levels of the hierarchy;
- there must be scope for expansion and development of the appraisal process as planning authorities gain more knowledge and experience of appraisal.

How to Perform Environmental Appraisal of Development Plans

There are a number of stages to environmental appraisal which allow it to be integrated into the development plan process, whilst ensuring that it is conducted in a systematic and comprehensive way.

Environmental Stock

As previously mentioned, appraisal needs a baseline position, so that policies and proposals are considered in terms of the changes that they are likely to make. Environmental stock provides that baseline. Policies can be examined to see if they improve or degrade the stock. Using environmental stock requires a prior attempt to catalogue all that has environmental value either in the plan area or, if it is part of a wider system, that environmental stock upon which actions in the plan area might impact (see English Nature 1992).

Due to the vastness of such a task, it is essential to recognise that identification of the stock and its condition can be undertaken in an incremental way. The authority will already have some information, and this can be supplemented as opportunities arise. A complete inventory of stock is not a prerequisite for appraisal. The task of appraising policies and proposals will identify gaps in knowledge and point to areas where environmental stock information needs to be developed. The task is to categorise the available information in a way that is useful for appraisal.

The use of environmental stock in the appraisal allows it to relate to something tangible and, to a limited degree, measurable. This enables:

- the systematic consideration of all environmental factors throughout the plan;
- consistency across planning areas, and with other environmental information bases held by different organisations;
- monitoring of the plan's effectiveness.

In drawing up the categories of stock there are three considerations:

1   identifying all significant aspects of the environment on which land use plans can have an impact;
2   distinguishing clearly between these so that their use in the appraisal will be informative;
3   keeping the number of components in the stock to as few as possible in order to keep the appraisal process manageable.

The research team was mildly surprised to find a broad level of agreement on the appropriate range of stock to be considered in environmental appraisal; an example being English Nature (1992). Three levels of concern were identified – global sustainability; natural resource management; and local environmental quality.

Global sustainability – the key component is the atmosphere; its composition and climatic stability. This is a response to concerns such as global warming and the erosion of the ozone layer (Department of the Environment 1992). The prime factor amenable to planning influence which is destabilising the atmosphere and leading to climatic change is fossil fuel use. Both energy in transport and energy in buildings are affected by planning policy. So is the rate at which fossil fuels are substituted by renewable energy sources. In addition, planning can affect absorption rates by protecting and promoting trees and woodland.

The environmental elements of most other aspects of stock can be more readily identified and described, and indicators established. For example, water is a key natural resource, and both its conservation and quality are of concern. Ground water levels and quality, together with river levels and quality, are indicators for the development plan. (A recommended list of elements of stock is provided in Table 21.1. Such a list can never be definitive, as it will need to change to suit the natural features of the plan area and as our knowledge of the impacts on the environment develops.)

Table 21.1   Recommended list of environmental stock

Level of Concern	Element of Stock

Global Sustainability	• Transport Energy Efficiency
	   Trip length
	   Number of trips
	   Public transport share
	   Walking and cycling share
	• Built Environment Energy
	   Efficiency
	   Heat loss
	   Capital energy
	   Potential for combined heat and power
	   Solar gain
	• Renewable Energy
	• Carbon Dioxide Fixing

Resource Management	• Air • Water • Land and Soil • Minerals • Wildlife
Local Environmental Quality	• Landscape • Liveability • Heritage • Open Space • Building Quality

Sources of information on environmental stock include the following:

- data held by the planning department on statutory designations, such as Areas of Outstanding Natural Beauty or National Parks;
- other departments of the same authority, including environmental health which may be the main source of data on air quality;
- external bodies such as the Ministry of Agriculture, Fisheries and Food which holds data on the quality of agricultural land, or the National Rivers Authority which has concerns for the protection of water resources;
- voluntary and local groups, perhaps wildlife trusts, which may be able to provide expert information on habitats in the plan area.

Where the data collected relates to a land area, it could be recorded on maps. However, a more sophisticated method is to use a computer based Geographic Information System (GIS), which not only allows the information to be stored, but eases its manipulation. GIS is particularly useful when it comes to cross-boundary information and grants authorities the ability to exchange and import environmental data. Again, although the GIS is useful, it is not a prerequisite for environmental appraisal.

The research team found that when collecting information on environmental stock, it may be beneficial to refer to specific targets or standards, which could be set by external bodies. Examples might include ground water concerns set by the National Rivers Authority, or air quality standards set by the World Health Organisation. Targets are also being set in the less charted areas of environmental concern, such as carbon dioxide emissions, and these could provide bench marks against which to judge policy impact.

Some of the case study authorities, and a number of other organisations, were attempting to define environmental capacities for certain areas. These may be based on estimates of impact on wildlife or mineral reserves, traffic levels and congestion, or open space. They therefore have no objective reality. Such capacities can have effects on neighbouring authorities, and need to be seen in context of wider regional planning priorities (see Kozlowski 1990). These locally defined environmental capacities should not, therefore, be incorporated into documents without public consultation and political debate. They are policy options which need to be balanced against social and economic aspirations.

Scoping the Development Plan

Scoping the plan is crucial to ensuring that the plan embraces the right scope of policies and proposals. This exercise, using an Environmental Checklist, may be performed alongside other scoping exercises but will be specifically aimed at guaranteeing that the range of environmental policies will move the plan towards greater sustainability. The Environmental Checklist has two stages:

1   defining the appropriate scope for the plan as set by government advice. A new

plan under preparation need only take this step;

2   checking the actual scope against the appropriate scope. Plans are often
       preceded by earlier versions (draft or adopted) and so the scope may need to be
       checked to ensure that all the appropriate issues are covered.

The prime source for the scoping exercise is the current advice on development plans.
In Britain this takes the form of Planning Policy Guidance notes which set out the
Government's policies on different aspects of planning. Issues of relevance to the
environment need to be brought to the fore. Against these, the scope of the plan can be
checked.

Stage two of the exercise is to compare the scope of the plan against the
intended scope. The key questions are: is the scope appropriate? If not, why not? The
recording of the answers must be systematic and explicit, noting where attention is
needed as the plan proceeds.

Scoping is not intended to be complex or time-consuming, but to provide a
more structured and rigorous framework for introducing the relevant scope of
environmental matters to the preparation of the plan than would otherwise be the case.
The scoping exercise offers the twin advantages of being both an inspiration and an
insurance, through:

- establishing consistency with central government policy and other relevant
  guidance;
- providing an early indication of the environmental issues which will require
  attention in the plan;
- putting the environmental agenda into the core part of the plan-making process;
- reducing the likelihood of challenges to the plan on grounds of inadequacy of
  the environmental content;
- drawing attention to potential alternative policy directions.

Appraising Plan Content

This might be considered as the core of environmental appraisal – identifying the
impacts and conflicts of policies and proposals, as well as highlighting alternative
options. Appraisal of the plan content is to be undertaken at a number of levels:

- when setting broad strategy objectives;
- at the formulation of the spatial strategy which broadly indicates how new
  growth is to be accommodated and linked to transport proposals;
- during policy development stage when policy options are being considered,
  chosen and refined;
- at site selection when land is being allocated for specific uses.

Stating broad objectives at the outset of plan preparation sets the context with which
subsequent policies and proposals should generally conform. Strategic objectives
should therefore support each other, even if only in general terms. Environmental

appraisal requires that if pursuit of an environmental benefit within one objective is undermined by the effect on another, then at the very least this should be made explicit. It can then be determined if the overall objectives are inherently incompatible or whether a different expression of the objectives can make them more compatible whilst still remaining true to their underlying aim.

At its simplest this can be a matrix with the strategy arranged along each axis. This creates cells in which broad consistency or inconsistency of each pair of objectives can be recorded.

Spatial strategies are generally returning to be an important part of development plans, partly in response to the desire to reduce carbon dioxide emissions by directing and managing travel demand through land use planning. It is particularly important that this kind of strategy is encompassed by the environmental appraisal, as noted by Barton and Stead (1993) and Owens (1986).

The interactions between transport and location are of such complexity that it may be advantageous to use some form of consistency analysis to guarantee that the overall strategy works. To achieve this within budgetary constraints and remain transparent requires the use of a compatibility matrix. Such a matrix arranges the strategy objectives along each axis and records the consequences of each pair of objectives in terms of being broadly consistent or inconsistent. For guidance on the use of compatibility matrices see Friend and Hickling (1987). Figure 21.1 and Table 21.2 display a worked example of a compatibility matrix together with an explanatory commentary.

The compatibility matrices must be simple to complete and understand – the following points will assist in this:

- use the matrix as a learning devise to pinpoint uncertainties or conflicts where further work is essential; complete it reasonably quickly;
- incorporate policy statements at different levels of generality as this assists in testing consistency between these levels;
- use a variety of symbols to record some degrees of conflict, reinforcement or uncertainty; compatibility does not necessarily imply mutual policy support, merely the absence of conflict;
- note the type of incompatibility, such as competition for scarce land resources or basic differences in purpose.

The advantages of undertaking a consistency analysis at the beginning of plan preparation are that it:

- tests compatibility between elements of the spatial strategy and key overall objectives and hence improves the effectiveness of the plan;
- provides an early indication of strategy objectives that may produce environmental conflicts;
- improves the environmental thrust of the framework within which detailed policies and proposals will emerge.

At the heart of the appraisal is the identification of the impact of each policy on each aspect of environmental stock. The policy impact analysis uses a policy impact matrix for this purpose, with the policies and the stock as the two axes. Each cell in the matrix confronts one policy with one aspect of the stock. The matrix is used to record whether there is an enhancing (positive), harmful (negative), or neutral impact.

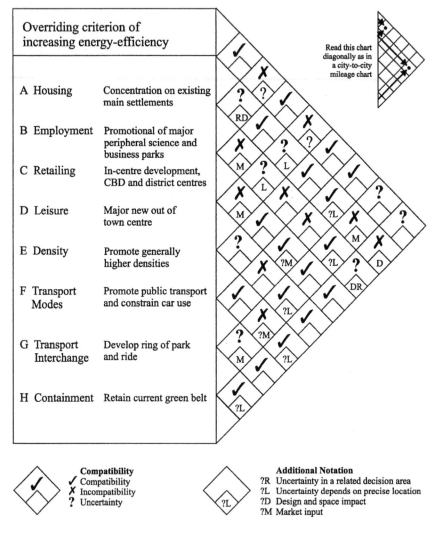

Figure 21.1 Compatibility matrix
Source: Department of the Environment 1993

Table 21.2   Compatibility matrix: commentary

	Comments	Action
STAGE 1. Identify the core of consistent policies		
A HOUSING CON-CENTRATION	Consistent	Incorporate in plan
C RETAILING IN-CENTRE	Consistent	Incorporate in plan
E DENSITIES HIGHER	Consistent	Incorporate in plan
F PROMOTE PUBLIC TRANSPORT	Generally consistent, but awkward in relation to retailing, because while the city centre is well served by public transport the district centres are not	Consider focusing any new retail development only on centres having public transport centrality
STAGE 2. Explore Uncertainties and modifications of politics that are not consistent		
B EMPLOYMENT (science and business parks)	Low density development in car dependant location is inconsistent with higher density public transport-based strategy and misses opportunities for mixed use, integrated development	Consider a compact 'science city' instead, combined with retail and leisure use, based on a public transport node <u>or</u> focus back on city centre
D LEISURE (new out of town centre)	Ex-urban single-use centre will have poor public transport accessibility	Link leisure centre to the 'science city' idea
G TRANSPORT INTERCHANGE (Park and Ride)	Park and Ride facilitates dispersal of population and low densities in the outer areas, undermining the viability of public transport in those areas	What realistic options are there? Consider Park and Ride as an <u>interim</u> policy. Aim for direct access to public transport services
H CONTAINMENT (green belt)	Green Belt constrains land available on the urban fringe and forces overspill to commuter settlements	Consider release of a sector of the green belt, related to growth around the science city or extend green belt to force concentration

Source: Department of the Environment 1993

The analysis can work in two ways:

1   an iterative way with the initial evaluation of impacts providing an indication of ways in which the policy needs to be refined;
2   it can provide a basis for considering relative performance when choices need to be made, for example when the draft plan is finalised.

For the policy impact matrix to remain understandable and useable, there are a number of helpful pointers.

- limit the size of the matrix to no more than 15 core elements of environmental stock;
- record the depletion or degradation of stock as a negative impact, and enhancement or protection as a positive effect;
- do not use any form of quantification – use only symbols;
- record only where the impact is clear and significant – do not represent a scale of impact;
- where an impact is likely, but unknown as it will arise from development of the policy in later proposals, this should be recorded, thereby flagging up the impact for future reference;
- impacts which are uncertain because of lack of knowledge must also be explicitly recorded and provide a focus for future work.

Such a matrix, when duly completed, will identify a number of relationships, including:

- groups of policies which perform well or badly against environmental stock;
- policies on one topic which appear to produce environmental effects contradictory to the intentions of other policies;
- apparent conflicts between different types of environmental consideration, such as those concerned with global sustainability and those with matters of local significance;
- areas where either the information base or the predictive techniques are inadequate.

Whilst the matrix is simple in concept, its completion is difficult, and it is recommended that this is undertaken by two or more people working together. These could be a plan team leader or an environmental scientist working in combination with individual topic specialists.

The complexity of the task may create the need for verification to ensure consistency and accuracy. Such verification can be performed in a number of ways, to suit the requirements of the circumstances. Appropriate means include review panels from within the planning department or with other departments, the involvement of other authorities, or the selective use of external bodies be they statutory consultees (such as the National Rivers Authority) or well informed pressure groups (such as the Council for the Protection of Rural England). This verification procedure must be integrated with the authority's own program of public consultation and must take account of confidentiality issues if there is any site specific material involved. It must be remembered that the task of arriving at a balanced judgement remains with the planning authority.

Policy Record Sheet

The complexity of the appraisal of policies, and the need for transparency demand that a record is kept of the appraisal at each stage. This includes a commentary on any significant factors which influenced that appraisal along with an actions column to remind the reader of the steps that need to be taken. Those policies with very little impact, or that are likely to be dropped from the plan may not require recording in this fashion, thereby reducing the time spent on the appraisal. The completed record sheets will constitute part of the explanation and justification for the inclusion of certain polices in the plan.

Presentation of the Appraisal

The appraisal needs to be transparent in use and an explanation of it should be part of the presentation of the plan. In this way the appraisal makes explicit how the environment has been taken into account in the formulation of the plan, and provides at least part of the justification for the policies and proposals set out in the plan.

In view of the pressure to keep development plans short and accessible (Department of the Environment 1992), it is suggested that the appraisal is presented in the following way:

- the plan should include a guide identifying which policies address each of the environmental issues to be covered by the plan;
- a chapter on the spatial strategy of the plan should be included, making explicit how the strategy is working towards greater sustainability in a consistent way;
- the environment chapter should include the aspects of environmental stock that have been used in the appraisal, and a summary of the approach that has been taken;
- each topic chapter should make explicit reference to the determining issues that have led to the chapter or policy as formulated;
- a separate report giving details of the appraisal process and summarising the findings should be prepared. As policies are challenged through the plan preparation process and its implementation, so the role of the appraisal will be challenged. A full statement will therefore be required so that the background work is available.

Expected Outcomes of Environmental Appraisal of Development Plans

The introduction of environmental appraisal will result in a number of benefits both in terms of the quality of the development plan and for its implementation and monitoring. These benefits include:

- a greater degree of compatibility between the strategy objectives, improving the effectiveness of the plan;
- a more sensitive and sustainable development plan with well defined and argued policies;
- the planning authority will be better placed when it comes to devising any amelioration schemes should development and environmental damage be unavoidable;
- the information gathered will provide a useful basis for any subsequent environmental impact assessments;
- a separate report giving full details of the background work involved in the environmental appraisal. As previously discussed, the separate report explaining the environmental appraisal in full is desirable as it helps to keep the development plan short and easy to read, whilst providing the opportunity to access the background work, such as the impact matrices and appraisal records;
- the baseline of environmental data can be monitored and readily updated, forming an up to date record of the environmental stock.

Implications for Development Plans and their Preparation

Environmental appraisal will, of course, have implications for the preparation of development plans, as well as the end result. The authority has to ensure that the full scope of environmental issues is covered in preparing and appraising the plan. The previously discussed scoping exercise will ensure this.

The appraisal will highlight where there is a lack of knowledge. This can be linked to the baseline of environmental data, pinpointing where and what kind of data is required. Any identified shortages of knowledge can be supplemented as opportunities arise, perhaps through links with other agencies.

The justifications for development and its subsequent impacts will have to be stronger and more well argued than might otherwise be the case.

Relationships beyond the Planning Department

Apart from the inter-departmental links established in the information and analysis stages of the appraisal, the process has implications for links outside the local authority. The requirement for appraisal puts the emphasis on discussion with other groups whose information and views may assist the appraisal. The range of outside agencies which may become involved is very wide, from national government bodies to locally based voluntary groups; from environmental pressure groups to parish and community councils. Consultation with such groups will be immensely valuable to the appraisal and to the quality of the final plan. The involvement of outside bodies helps to validate the environmental appraisal and ensure that environmental interests are properly accounted for.

Perhaps the most valuable contribution of outside bodies is in assisting with building the environmental baseline. The range of data collected and monitored by such groups can be drawn upon by the local authority to keep its own baseline up to date.

The open approach promoted adds impetus to the participatory approach to planning for sustainability advocated by Local Agenda 21 – a product of the Rio Earth Summit. Environmental appraisal will inevitably increase the level of involvement, if initially on the planning authority's terms. This supports the commitment of those countries which signed Agenda 21 in Rio to achieve the objective of sustainable development and growth. Perhaps it is more appropriate, in the short term, to suggest that environmental appraisal in the preparation of development plans will help to reduce the incidence of unsustainable development, albeit a long and incremental process.

In terms of environmental protection, environmental appraisal (if performed by all planning authorities) contributes to the broader level of strategic environmental assessment at regional, national and international levels, helping to identify key areas of concern and establishing a useful bank of information, as noted by Therivel (1992) and Wood (1992). Similarly, appraisal will aid the preparation of environmental impact assessments for large projects, pinpointing where such schemes would be inappropriate and highlighting the environmental issues to be covered by the assessment.

Effective Land Use Planning

Environmental appraisal of development plans could be seen as crucial to effective land use planning and a major contribution to resolving the conflict between environmental protection and development. By explicitly bringing to the fore the potential impacts of development plans and policies, appraisal contributes to the political process of balancing economic growth, society and the environment in a systematic and transparent way.

Furthering Environmental Appraisal of Development Plans

Environmental appraisal of plans is a newly emerging activity and will be a rapidly changing field as new knowledge is fed into environmental planning. In particular there is likely to be significant development in the use of targets for measuring success of policies and emerging policy appraisal is likely to combine targets with matrices.

There is a need to monitor the effects of environmental appraisal on development plans. Will the new plans prove to be more sensitive to the environment? Or will the established body of social and economic pressures outweigh the current concerns with the protection of the environment?

As planning authorities take on the challenge and try to find the most efficient and manageable methods of appraising their plans, opportunities to research the developments of the appraisal process will arise. These could lead to some very worthwhile research projects, further refining the appraisal process.

Note

1   Neil Border is working at the Faculty of the Built Environment, University of the West of England, Bristol, England.

References

Barton, H., and D. Stead (1993) *Sustainable Transport for Bristol: Research paper*, University of the West of England, Bristol, UK.
Beanlands, G. (1990) *Scoping Methods and Baseline Studies in Environmental Impact Assessment*, Unwin Hyman Ltd., London.
Breheny, M. (1992) *Sustainable Development and Urban Form*, Pion Ltd., London.
British Government (1994) *Sustainable Development – The UK Strategy*, HMSO, London.
Clark, M., and Herrington (eds) (1988) *The Role of Environmental Impact Assessment in the Planning Process*, Mansell Publishing Ltd., London.
Commission of the European Communities (1991) *Draft Proposal for Directive on the Environmental Assessment of Policies, Plans and Programmes*, European Community, Luxembourg.
Department of the Environment (1992) *Development Plans: A good practice guide*, HMSO, London.
Department of the Environment (1992) *Planning Policy Guidance Note 12: Development plans and regional planning guidance*, HMSO, London.
Department of the Environment (1992) *Land Use Planning Policy and Climate Change*, HMSO, London.
Department of the Environment (1993) *The Environmental Appraisal of Development Plans: Good practice guidelines for local planning authorities*, HMSO, London.
English Nature (1992) Strategic Planning and Sustainable Development: An Informal Consultation Paper, *English Nature*, Peterborough, UK.
Friend, J., and A. Hickling (1987) *Planning Under Pressure: The strategic choice approach*, Pergamon Press, Oxford, UK.
Jacobs, M. (1991) *The Green Economy*, Pluto Press, London.
Kozlowski, J.M. (1990) Sustainable Development in Professional Planning: A Potential Contribution of the EIA and UET Concepts, *Landscape and Urban Planning*, No.19, 1990, pp307-332.
Owens, S. (1986) *Energy, Planning and Urban Form*, Pion, London.
Therivel, R., et al (1992) *Strategic Environmental Assessment*, Earthscan Publications Ltd., London.
United Nations Commission on Environment and Development (1987) *Our Common Future* (Brundtland Commission), United Nations, New York, NY.
Wood, C. (1992) Antipodes: Strategic Environmental Assessment in Australia and New Zealand, *Project Appraisal*, Vol.7, No.3, September 1992.

Chapter 22

Integrated Environmental Zoning in Local Land Use Plans: Some Dutch Experiences

H. Borst[1]

22.1 Introduction

In The Netherlands, integration of policy fields is a trend which promises to bring together spatial planning and environmental policy. There is a growing consciousness of the connection between these two policy fields. Both spatial planning and environmental policy deal with the quality of the environment of human beings. An effectively functioning physical structure as well as, for instance, clean air determines the quality of the total environment.

Especially in urban areas, there can be a conflict between activities, such as manufacturing, which reduce environmental quality and activities that are sensitive for environmental pollution, such as residential areas. From the perspective of spatial planning, the locations of these activities should be close to each other, but from the perspective of environmental policy these activities have to be separated from each other (de Roo 1993). The task is to combine these opposite interests into an integrated policy for the physical and natural environment.

One example of the integration of spatial planning and environmental policy is the Dutch system for Integrated Environmental Zoning (IEZ). The purpose of this system is to create desirable spatial and environmental quality in the vicinity of complex industrial activities. Some experiments with the application of this system for integrated environmental zoning have demonstrated that there are several problems to be solved. One of the major problems is the incorporation and implementation of the system for integrated environmental zoning into the practice of spatial planning.

The focus of this chapter is the integration of the system for IEZ in Dutch spatial planning at the local level. There are several obstacles for this integration caused by opposing interests. For instance: from the spatial perspective a new residential area is needed in a part of a city, but from the environmental point of view this residential area can not be built because of the high environmental load (level of pollution) from industrial activities. The central question in this chapter is how IEZ can be procedurally integrated with Dutch spatial planning, in order to cope with this tension between the spatial planning and the environmental policy.

The next two sections provide a brief reflection on the Dutch system for integrated environmental zoning and on the Dutch system for spatial planning at the local level. Section four deals with some practical experiences with the integration of spatial planning and environmental policy in the framework of the IEZ system. The last section of this chapter contains some suggestions to improve the incorporation and implementation of integrated environmental zoning in spatial planning policy.

22.2   Integrated Environmental Zoning, a Brief Description

In 1990 the Ministry of Housing, Spatial Planning and Environment introduced a system for integrated environmental zoning (VROM 1990). This system is a provisional one and is intended 'as a tool to help in determining the cumulative environmental load and in establishing the integral environmental zone in pilot projects' (VROM 1990). This system will be applied in projects where complex industrial installations produce environmental impacts which influence the environmental quality in near-by residential areas. The following categories of environmental loads are included in this system: industrial noise, odour, major industrial hazard and local air pollution caused by toxic or carcinogenic substances.

The provisional system for integrated environmental zoning provides a technique with which measures of these different kinds of environmental loads can be added together. The different types of sectoral environmental loads are translated to one integrated environmental load. Because of the distance-related character of these environmental loads (with a growing distance from the source the environmental pollution will decrease) the integrated environmental load can be expressed in spatial zones. These integrated environmental zones can be labelled as ranging from 'clean areas' to 'severely polluted areas'. It is possible to connect spatial consequences to these environmental zones. In the Dutch provisional system these spatial consequences vary from 'no consequences' for the clean areas, to 'pulling down of houses' and 'no residential development' for severely polluted areas. The process of integrated environmental zoning consists of three stages:

1   investigation of the environmental situation and the spatial structure;
2   administrative decision making; and
3   implementation and execution of the results.

The first stage in IEZ includes measuring the integrated environmental load for an impacted area and a survey of the spatial structure and of the spatial plans for the impacted area. Based on this information, the administration makes decisions concerning the location of environmentally intrusive activities and environmentally sensitive land uses. These decisions determine the location of the integrated environmental zones.

There often exists a difference between the actual and the desired integrated environmental zones. In these situations it is theoretically possible to reduce pollution at the source, or to execute effect-oriented measures in the spatial sphere (for instance restrictions in land use). The different types of source-measures and effect-oriented measures have to be weighed one against another. Because the implications of pollution reduction at the source are less dramatic than the execution of effect-oriented spatial measures, the first type of measures will be preferred. The task for the administration is to create a balance between the environmentally intrusive activities and the environmentally sensitive functions.

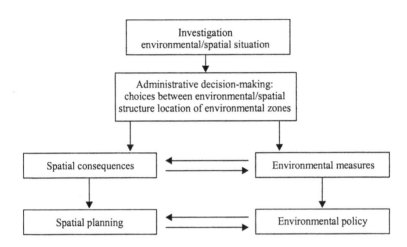

Figure 22.1   Process of environmental zoning

In the last stage, the spatial restrictions on use of polluted areas and the pollution abatement measures have to be implemented. The possible restrictions in land use have to be incorporated into the local land use plans. There can be a strong relation between the spatial consequences and environmental measures. For example, residential development is possible if the environmental load is reduced.

22.3   Local Land Use Planning

The Dutch system for local land use planning is characterised by two types of spatial plans; a strategic plan and a more operational one. The strategic plan (*structuurplan*) contains the goals and major outline of spatial development for the whole area of the municipality. An important function of this strategic plan is the integration of different policy fields at the local level. The spatial implications of environmental policy can be included in this plan. In this way a local administration can create a comprehensive vision of the general spatial development of the municipality.

This comprehensive vision can be the basis for the more operational plan (in The Netherlands this plan is called the *bestemmingsplan*). The purpose of this plan is to regulate the land use. Usually the spatial scale of this plan will vary from one single site to a whole neighbourhood or an area with a specific function such as industry or residence. In contrast to the strategic plan, this plan is legally binding for those who will use the land. Land use inconsistent with this plan is prohibited.

One of the practical problems is the implementation of environmental themes in this planning system. In the next section we will discuss two examples of this implementation process.

Strategic level
- global vision on spatial development
- integration of different policy fields

Operational level
- detailed spatial scale level
- legal regulation of land use

Figure 22.2   Two levels of local land use planning in The Netherlands

22.4   Relation between IEZ and Spatial Planning in Practice

The provisional system for integrated environmental zoning is being applied in several pilot projects. Most of these projects have finished the investigation of the integrated environmental load. The results of this investigation have been compared with the actual and desired spatial structure. In some projects the results of this investigation and comparison have great implications for the spatial planning. There are situations where, because of the consequences of the environmental zones, pursuing a spatial policy is difficult (Borst and de Roo 1993). In these cases it is also impossible to achieve a straight implementation of the environmental zones in the local land use plans. It is necessary to create a procedural means to integrate spatial and environmental policy that can cope with these problems.

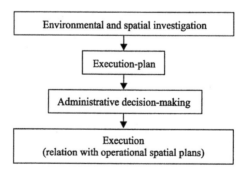

Figure 22.3   IEZ process in Arnhem

In the pilot project in Arnhem, the central element is an industrial area located close to the inner city. The application of the provisional system for IEZ creates a situation in which the whole city of Arnhem falls into the category of severely polluted areas (Boei 1993). This predetermines the spatial policy of the city of Arnhem. It is not clear whether it is possible to take adequate source measures to reduce the environmental pollution from the industrial area. Consequently this project resulted in an impasse because it is not reasonable to suppose that the enormous spatial consequences, such as no new development and even removal of much of the housing stock, can be implemented in the spatial policy. It is clear that there are no short-term solutions to these problems in Arnhem.

In the IEZ project in the city of Maastricht (Borst 1993), the purpose was to improve the environmental and spatial quality of three selected areas. Because these areas are impacted by a high level of several kinds of pollution and because they are deteriorated areas, are being considered in the Strategic Plan for redevelopment. Both the environmental pollution and physical quality problems are considered together, making the procedure in Maastricht different than that used in Arnhem.

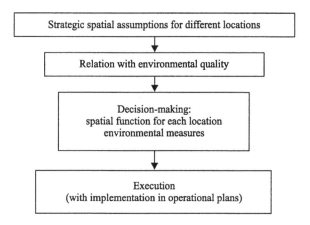

Figure 22.4   IEZ process in Maastricht

An important feature of the procedure in Maastricht is the use of strategic spatial assumptions for different locations within the three selected areas. For each location a number of possible land use options are considered (Maastricht 1989). These spatial options are then related to the environmental quality of the specific location. The next step is to make choices concerning the specific functions (land uses) for each of the locations in relation with the integrated environmental quality (Maastricht 1993). If the spatial function is sensitive to high environmental load, as are residential areas, this environmental load has to be reduced through greater control of pollution at the sources. On the other hand, if the spatial function is less sensitive to environmental load, there will be more environmental space for industrial activity. In this procedure there is a direct relationship between the spatial functions and the environmental quality.

Thus in practice, there is a difference in the procedures employed for integrated environmental zoning in these two cases. In Arnhem the IEZ process is rather independent from the spatial planning policy; the procedure provides a coupling between IEZ and the spatial planning at an operational level. The environmental problem in this project requires a solution process beginning at the strategic level, after which strategic choices can be implemented in the operational plans.

In Maastricht the solution process begins at the strategic level. In an early stage of the process, the spatial and environmental issues compared with each other. This procedure anticipates on a situation in which the spatial policy, at an operational level, will have to deal with an unacceptably high environmental load. A second advantage of this approach is the possibility for the concerned parties to participate in the process of integrated environmental zoning. In this way it is easier to create a basis for the acceptance of the consequences of the IEZ process.

22.5   A Framework for Integration

Application of integrated environmental zoning, without taking the spatial planning policy into account, may cause a troublesome situation. Because of insufficient procedural co-ordination, co-ordination of substantive concerns will also be a problem. In these situations, an IEZ project may well come to an impasse. Especially in complex situations where the conflicts between environmentally sensitive land uses and environmentally intrusive activities are extensive, an integration between IEZ and spatial planning at an operational level is not sufficient. In these situations it is necessary for spatial policy to consider and deal with the integrated environmental load in an earlier stage. A more detailed solution of the conflicts between the environmental quality conditions and the spatial structure can be implemented at the level of operational spatial planning.

In this framework integration of spatial and environmental issues takes place at both the strategic and operational levels. At the *strategic level* there is a general assessment of the implications of the integrated environmental load for the spatial structure and spatial plans. This environmental quality information can be an important basis for planning land uses, for instance new residential areas, in appropriate locations. New locations may need to be found for environmentally sensitive activities, and future reduction of pollution may make currently inappropriate locations for these activities acceptable for them after abatement has taken place.

The next step is to make choices concerning the environmental load in relation to the spatial structure at the *operational level*. As a result of these choices the location of the integrated environmental zones will be determined. The consequences of these choices (spatial consequences and/or environmental measures) can then be implemented by means of the local environmental policy and the operational spatial planning.

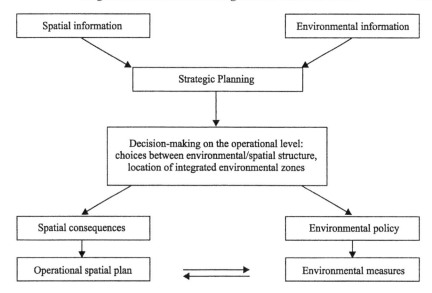

Figure 22.5   A framework for integration

At this operational level, there is an interdependency between the environmental load and the spatial structure/plans. In most cases, the environmental load can not be reduced in a short period of time. Another problem can be the existence of environmentally sensitive urban activities in areas with a high integrated environmental load. It is drastic action either to pull down whole residential areas or to remove industrial activities. In these sorts of situations, it has to be possible to introduce a form of spatial or temporal flexibility in the operational land use plan.

A flexible land use plan can be useful in situations calling for new spatial developments. The operational land use plan will take into account the environmental load at various locations. Until the desired environmental quality has been achieved, further spatial developments are restricted. After the reduction of an environmental load, the spatial function of an area can be changed to a more environmental sensitive land use. When the land use plan can anticipate future environmental improvement, the existing conflict between environmentally intrusive activities and environmentally sensitive land uses will be tolerated for a given period of time.

22.6   Conclusion

To solve conflicts between environmental policy and plans for new urban development, these two activities need to be co-ordinated. This integration should begin at the strategic planning level, since this way it is possible to find solutions at a more general level both in terms of alternative locations for new development and the timing of the development. These decisions then can be worked out in detail at the operational planning level. Using this framework, it is possible to cope with the problems between environmentally conflicting activities in urban areas.

Note

1　　　Hermen Borst is working for the Directorate General for Spatial Planning (*DG Ruimte*), Dutch Ministry of Housing, Spatial Planning and Environment.

References

Boei, P.J. (1993) Integrale milieuzonering op en rond het industrieterrein Arnhem-Noord [Integrated environmental zoning and the industrial area of Arnhem-North], in G. de Roo (ed.) *Kwaliteit van norm en zone* [Quality of the environmental standard and zone], Geo Pers, Groningen, The Netherlands.

Borst, H., and G. de Roo (1993) Integrale milieuzonering en de ontketening van de ruimtelijke ordening [Integrated environmental zoning and the unleash of the spatial planning], *Planologische diskussiebijdragen*, Delft, The Netherlands.

Borst, H. (1993) *Integrale milieuzonering en ruimtelijke ordening* [Integrated environmental zoning and spatial planning], Faculty of spatial sciences, University of Groningen, Groningen, The Netherlands.

Maastricht [municipality] (1989) *Project Integratie Milieubeleid Maastricht, Structuurbepalende aannamen. Ruimtelijke alternatieven voor de drie proefgebieden Boschpoort, Limmel en Groot Wyck* [Project Integration Environmental Policy Maastricht. Spatial alternatives for the three experimental areas Boschpoort, Limmel and Groot Wyck], Maastricht, The Netherlands.

Maastricht [municipality] (1993) *Eindrapportage PISA, ruimtelijke keuzes en milieusaneringen* [Final report PISA, spatial choices and environmental redevelopment], Maastricht, The Netherlands.

Roo, G. de (1993) (Integrale) Milieuzonering: Patstelling of Synthese?! [(Integrated) Environmental Zoning: Impasse or Synthesis?!], in G. de Roo (ed.) *Kwaliteit van norm en zone* [Quality of the environmental standard and zone], Geo Pers, Groningen, The Netherlands.

VROM [Dutch Ministry of Housing, Spatial Planning and Environment] (1990) *Ministerial manual for a provisional system of integral environmental zoning*, IMZ-reeks deel 14, The Hague.

Chapter 23

Integrating Environmental Quality Criteria in Municipal Master Plans: A General Proposal

C. Garrett[1]

Summary

The growing importance of environmental problems and their intimate relationship with land use changes require the integration of preventive and effective environmental management practices in land use planning. This chapter presents a set of methodological proposals for the assessment of environmental quality in terms of the main pollution types – air, noise, water and soil – and a model of sectoral environmental zoning, to be integrated in the preparation of municipal land use plans (master plans), as a result of research carried out at the Technical University of Lisbon.

The general institutional systems for land use planning and environmental management in Portugal are presented, in order to provide the framework for the proposals that have been developed. The proposed methodology comprises a model for the preliminary evaluation of environmental quality problems at local level, and four analysis models – for air, noise, water and soil pollution – that result in a form of environmental zoning to be applied in the preparation of land use master plans. An outline of further development of this research is also presented.

23.1 Approaches and Underlying Concepts

Regional and urban planning and environmental protection policies have been tracking parallel courses for a long time, converging to a common goal – improving the human environment. This fact is acknowledged not only in the literature concerned with the evolution of these two branches of science and policy (Lichfield and Marinov 1977; Lovejoy 1979; Huang 1990; Kozlowski 1990), but also in documents devoted to the big concerns and ways to resolve the most important environmental problems at global level (IUCN 1980; CMAD 1991).

In Portugal, a balanced environment is one of the constitutional rights and obligations of the citizens, and the Environment Act assumes that the goal of environmental policy is to optimise and assure the qualitative and quantitative perpetuity of the use of natural resources, as a basic prerequisite for a self-sustained development. It also considers that integrated land use planning at regional and municipal levels is a fundamental instrument for the enforcement of environmental policies, in this way recognising the intimate relationship between both areas.

Planning as a rational attitude for the use of land and its resources is a condition for the promotion of a sustainable development model. 'Indeed, it is only by means of planning, prediction, monitoring and the use of controls, either voluntary or statutory, that the destructive processes of exploitation (of natural resources) may be halted' (Lovejoy 1979).

The Role of Planning in Pollution Prevention

Land use transformation processes are certainly responsible for a series of environmental pollution problems, which can not be solved only at the project assessment level (Partidário 1992). Wood (1976) states that 'the most powerful contribution' of planning authorities 'is at the first of the stages in determining the nature and location of new development and redevelopment'.

Controls exercised at later stages of the pollution process (source-oriented or effect-oriented measures) are less effective in preventing pollution problems, or will at least respond in less comprehensive ways to the existing or future pollution problems, and do not integrate scarce resource management principles. The decision-making context of planning is the only process adapted to an anticipative consideration of the effects of cumulative pollution charges.

Environmental legislation and management aims at reducing pollution as much as possible, through the use of Best Available Technical Means – source-oriented measures. It also seeks to assure public health, human well-being and the conservation of natural systems and values, therefore requiring natural resources management controls – effect-oriented measures.

Planning can have an important role in this context 'as the location of potentially pollutant activities is determined and as development may be controlled in the proximity of pollution sources' (UK Government 1992). Its action does not focus on pollution control, but rather on:

- development;
- the impact of development in the quality of life;
- the potential of risk or contamination;
- preventing nuisance;
- controlling the impact of roads through changes in traffic levels, etc.

The conditions posed by planning can therefore complement the pollution control and natural resources management regimes.

23.2 Integrating Pollution Problems in Planning: A Proposal

Objectives of the Research

One component that has been left aside in Portuguese urban and municipal planning practice is pollution. The work carried out within the framework of a MSc thesis (Garrett 1993) was aimed to demonstrate the feasibility and the advantages of introducing an analysis of pollution problems in the preparation of plans. This analysis should not require, at least at a preliminary stage of plan preparation, technical expertise and environmental data that are not yet available at most of the 305 Portuguese municipalities.

The proposal consists of a set of four sectoral analysis models for air, water, soil and noise pollution, and a preliminary scoping procedure for existing or expectable pollution problems in a municipality. Two basic objectives inform this proposal:

- integrating environmental quality concerns and objectives in the process of preparation of land use plans;
- integrating environmental objectives and environmental management instruments in land use management policies.

Levels and Scales

Although in operational terms urban planning is the most adapted level for defining environmentally compatible land use distribution, at the municipal master plans 2 level (1:25,000 scale) a preliminary assessment of environmental quality and existing or potential pollution problems can be useful to improve the quality of the plans themselves, in order to better correspond to general goals of environmental protection and enhancement. Municipal master plans intervene in land use changes at a scale and scope which are the framework of development projects, thereby acting upon the decisions which influence environmental quality in a more strategic way.

The municipal master plans correspond to the strategic level of planning at which the most stable and structuring options on land use development are considered, and while focusing on the middle term do this by taking long term goals and possibilities into account (Lobo et al. 1990). Besides, the scale of a municipality 'presents itself as the most adequate level to a balanced exercise of land use (and environmental) management, because the reasons of political and administrative power are nearest to real people and to the real places where they live' (Pardal 1988).

Roles and Actors

The integration of pollution matters in plans is a general principle in Portuguese law but it is not a direct obligation upon plans. Environmental quality concerns are more binding at project level than at the strategic planning level. Figure 23.1 illustrates the presence of requirements about environmental pollution in the legislation on planning (at global, sectoral or project level), and Figure 23.2 shows how, inversely, environmental legislation domains are intimately related with planning activities of various kinds.

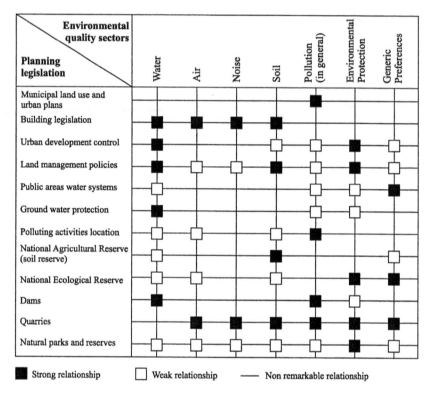

Planning legislation \ Environmental quality sectors	Water	Air	Noise	Soil	Pollution (in general)	Environmental Protection	Generic Preferences
Municipal land use and urban plans					■		
Building legislation	■	■	■	■			
Urban development control	■				□	■	□
Land management policies	■	□	□	■		■	□
Public areas water systems	□				□		■
Ground water protection	■					□	
Polluting activities location	□	□			■		
National Agricultural Reserve (soil reserve)	□			■			□
National Ecological Reserve	□	□		□		■	■
Dams	■				■	□	
Quarries		■	■	■	■	■	■
Natural parks and reserves	□	□	□	□		■	□

■ Strong relationship　　　□ Weak relationship　　　— Non remarkable relationship

Figure 23.1　Environmental requirements in the Portuguese legislation on land use planning

Description of the Proposals

Model for the preliminary scoping of environmental problems and objectives The model proposed in this research for the preliminary scoping of existing or expectable pollution problems in a municipality is largely inspired by the 'Environmental Advisory Service' developed by White et al. (1985). In both cases fast and efficient approaches to the identification of existing or expectable environmental problems are sought, while not requiring a specialised knowledge in environmental assessment – although a practical knowledge of the field is needed – and acknowledging environmental issues as one of the components of land use planning.

For the application of the model, a form was prepared with a set of questions to be put to the local planner in order to lead him/her to a systematic survey of the main factors of environmental pollution present in his/her municipality, according to a checklist approach. These factors should be portrayed by location on a map and the pollution intensity qualitatively defined, this way drawing a preliminary image of the effective or potential risk of degradation of the environment in the area.

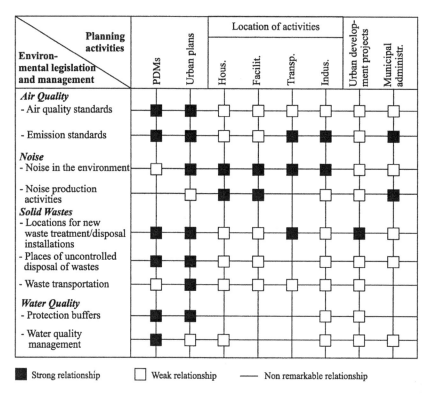

Figure 23.2 Relationships between environmental quality legislation and planning activities

The result is a global or city-wide assessment, by categories and sub-categories of questions, for the whole of the municipality and for some sub-areas, allowing the identification of:

- a general topology and aggregation of environmental quality problems in the municipality and its sub-areas – including more important, less important, improbable or unknown problems;
- a relative importance of the several categories of questions – what factors are more important to the environmental pressures recognised in the area.

These results should help define:

- the level of importance of the environmental objectives to be considered by the plan to be prepared;
- the general potentials or limitations placed on the general development of the area by pollution problems;
- the areas expected to be more exposed or more vulnerable to environmental quality problems, to be focused on by more detailed studies during the preparation of the plan.

Sectoral models Natural resources are inputs to land use, as they are necessary to the development of human activities, but they also suffer the effects from the pollution resulting from those activities.

The problem of planning in this context is to determine the level to which land uses can be developed without exceeding the maximum load the natural systems can support without disruption, that is, not jeopardising the sustainability of global, regional and local ecosystems. Within these absolute limits, planning must define where the land uses are most efficient (in the use of natural resources and in implementing development objectives) and what characteristics for each land use in each location may be acceptable. This way, good decisions involve not only location or zoning options but also management programs.

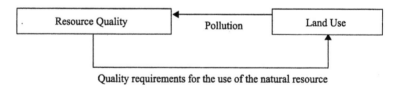

Quality requirements for the use of the natural resource

Figure 23.3   Relationship between the quality of natural resources and land uses

For each domain considered in this study (air, water, noise and soil), a sectoral environmental zoning procedure was designed, following four steps:

1   identification of polluting land uses and assessment of their relative pressure, through potential pollution risk indices;
2   identification of quality objectives through the identification of levels of sensitiveness of the existing land uses/land systems;
3   identification of areas and levels of potential conflict between existing land use and environmental quality, to be considered in the land use strategies of the plan (environmental zoning);
4   definition of policies to manage potential and existing conflicts.

The strategic component of plan design can be supported by:

- the simulation of several situations of land use (location, types and intensities);
- the simulation of several situations of resource availability and quality, depending upon alternative management policies.

The general design of the proposal is illustrated by an assumed municipality, comprising a set of resources and land uses represented in Figure 23.4, and with the analysis of water pollution-related problems.

Legend:
- urban (dispersed)
- urban (dense)
- industry
- forest
- agriculture
- coastal area
- water system

Figure 23.4 Land use map

First of all, an identification of water resources and water uses is necessary (Figure 23.5). Through a typification of land uses according to the water pollution risks they represent (Fabos 1985), it is possible to determine the areas where the pollution risk is more important, considering the location of these land uses (as pollution sources) and the sense and pattern of drainage (Figure 23.6).

Simultaneously, water uses have specific quality requirements; in a simplified classification, we can consider, in a rank order of decreasing water quality requirements:

1   public supply, fish or conchological breeding and some specific industrial supplies;
2   irrigation, recreation with direct contact and fishing;
3   recreation without direct contact and energy production.

Comparing the spatial distribution of the classes of pollution risk with the classes of water quality requirements allows an identification of potential conflict situations between land uses and water uses (Figure 23.7).

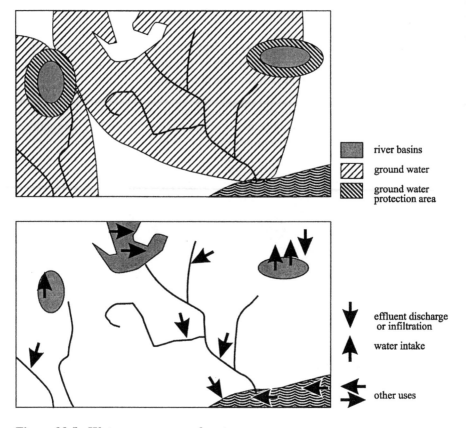

Figure 23.5   Water resources and water uses

For each type of zone (and based on more detailed analysis, for each area), measures to reduce pollution can be defined, besides helping to decide about the location of new uses in the municipality. Examples of such measures may be:

- limits put to the occupation of aquifer recharge areas;
- restrictions to further industry development where pollution already is unacceptable;
- limits to urban growth;
- monitoring network location;
- areas suitable for tourism development, etc.

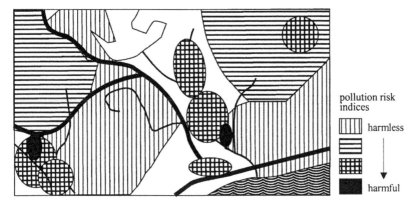

Figure 23.6   Water pollution risk indices

This approach can be applied in greater detail and prescriptiveness by using quantitative analysis, where water-polluting land uses are characterised in terms of their specific type of pollutant emissions, or even by locating the exact spots where discharges occur and modelling their dispersion, and water quality and vulnerability indices can be determined by using water quality data and water uses specific water quality requirements.

The definition of air pollution risk indices must consider not only the location of pollutant sources by land use (different types of industry, airports, major traffic ways and terminals), but also a pattern of pollution dispersion at the regional level or, at least, the consideration of a fixed radius of influence of each type of polluting land use.

		Pollution risk indices		
		1	2	3
Quality requirement indices	1	Black areas: very probable conflict between land and water uses		
	2	Grey areas: probable conflict between land and water uses		
	3	White areas: unlikely conflicts between land and water uses		

Figure 23.7   Definition of criteria for the definition of a water quality sectoral zoning

A grid approach may be more appropriate for the cartographic representation if a type of neighbouring analysis is adopted. The areas of potential conflict between land use and air quality are assessed by defining which land uses are more vulnerable to the effects of air pollution.

For noise, the proposed method is an example of a qualitative approach to nuisance problems. Considering those land uses which are probable sources considering the levels of noise attenuation of different types of land use, according to their physical characteristics, probable situations of the quality of the environment in terms of noise can be estimated. The steps for the analysis are:

• outline areas which constitute direct noise emission areas;
• outline areas with similar noise attenuation characteristics;
• estimate noise in the environment and therefore areas subject to different noise nuisance risk.

Figure 23.8   Water quality sectoral zoning

For soil contamination, the same logic is applied: identifying land uses and activities which potentially cause soil contamination and identifying the areas where natural characteristics or other land uses are vulnerable to such situations helps in identifying the problem areas (where those situations are most probably existing) and the different areas where moderate, low or unknown risk may occur.

23.3   Further Development

The application of these proposals to a municipal master plan is under preparation; it is intended to use a plan which is practically ready or already in force and assess if the development of an analysis of these four domains bring some new ideas to the plan and to its management policies. Following the experience that has been gained from the METLAND project, at the University of Massachusetts (Fabos 1985), and that has been used in several planning projects in the USA and also in Europe, an application of these principles with the support of a Geographic Information System and associated computer spatial modelling tools is envisaged as a medium term project, through the development of a quantitative approach.

Environmental sectoral zoning seems to be an applicable solution to the integration of environmental quality concerns in the preparation of land use plans. The process of integrating the results of such sectoral analysis must be defined in each case, according to the plan methodology employed.

The improvement of the quality of municipal plans in the sense of greater integration of environmental quality concerns and sustainable development goals is presently limited by the scarce technical means and limited environmental data available to local administrators, and also by the low priority accorded to environmental quality aspects in the framework of local management.

Note

1   Cristina Garrett is working at the General Directorate of the Ambiente of Lisbon, Portugal.

References

CMAD (Comisso Mundial sobre Ambiente e Desenvolvimento) (1991) *O Nosso Futuro Comum*, Meribrica/Liber, Lisbon, Portugal.

Fabos, J.G. (1985) *Land-Use Planning: From Global to Local Challenge*, A Dowden & Culver Book, Chapman and Hall, New York, London.

Garrett, C. (1993) *A integrao de critrios de qualidade do ambiente na elaborao de Planos Directores Municipais*, master's thesis, Universidade Técnica de Lisboa, Lisbon, Portugal, (n/p).

Ghiglione, R., and B. Hatalon (1978) *Les Enqutes Sociologiques, Theories et Pratique*, Armand Colin, Paris.

Huang, S. (1990) A Land-Use Suitability Approach for Integrating Impact Assessment with Development Planning, *Impact Assessment Review*, Vol. 8, No. 1-2, pp. 233-247.

IUCN (1980) *Estratgia Mundial de Conservao*, Serv. Estudos do Ambiente, SEUA, Lisbon, Portugal.

Kozlowski, J. (1990) Sustainable Development in Professional Planning: A Potential Contribution of the EIA and UET concepts, *Landscape and Urban Planning*, Vol. 19, pp. 307-332.

Kozlowski, J. (1985) Threshold Approach in Environmental Planning. *Ekistics*, Vol. 311, pp. 146-153.

Lichfield, N., and U. Marinov (1977) Land-Use Planning and Environmental Protection: Convergence or divergence?, *Environment and Planning A*, Vol. 9, No. 9, pp. 985-1002.

Lobo, M.C., et al. (1990) *Normas Urbansticas*, Volume 1 - Princpios e Conceitos Fundamentais, DGOT/UTL, Lisbon, Portugal.

Lovejoy, D. (ed.) (1979) *Land Use and Landscape Planning - 2nd. Edition*, Leonard Hill, Glasgow, UK.

McHarg, I. (1967) *Design with Nature*, The Natural History Press, New York, NY.

Pardal, S.C. (1988) *Planeamento do Territrio – Instrumentos para a Anlise Fsica*, Livros Horizonte, Lisbon, Portugal.

Partidário, M.R. (1992) *An Environmental Assessment and Review (EAR) Procedure – A Contribution to Comprehensive Land-Use Planning*, A thesis presented for the degree of Doctor of Philosophy, August 1992, University of Aberdeen, Aberdeen, UK, (n/p).

Petak, W.J. (1981) Environmental Management: A system approach, *Environmental Management*, Vol. 5, No. 3, pp. 213-224.

UK Government (1992) *Planning Policy Guidance Note: Planning and pollution control* (Consultation draft), 1992, n/p.

VROM [Dutch Ministry of Housing, Physical Planning and Environment] (1990) *Ministerial Manual for a Provisional System of Integral Environmental Zoning/Integrale Milieu Zonering*, VROM, The Hague, The Netherlands.

White, K.P., et al. (1985) The Environmental Advisory Service (EASe): A Decision Support System for Comprehensive Screening of Local Land-Use Development Proposals and Comparative Evaluation of Land-Use Plans. *Environment and Planning B: Planning and Design*, Vol. 12, pp. 221-234.

Wood, C.M. (1976) *Town Planning and Pollution Control*, Manchester University Press, Manchester, UK.

Index

Milton Keynes UK
Ingram Content Group UK Ltd.
UKHW031128141024
449569UK00006B/347